P9-BZE-854

DOLLARS AND SENSE

ALSO BY DAN ARIELY AND JEFF KREISLER

By Dan Ariely

Predictably Irrational

The Upside of Irrationality

The (Honest) Truth About Dishonesty

Irrationally Yours

Payoff

By Jeff Kreisler

Get Rich Cheating

DOLLARS AND SENSE

HOW WE MISTHINK MONEY AND
HOW TO SPEND SMARTER

DAN ARIELY
AND JEFF KREISLER
WITH ILLUSTRATIONS BY MATT TROWER

HARPER
An Imprint of HarperCollins*Publishers*

This book is designed to provide readers with a general overview of financial thinking and how it works. It is not designed to be a definitive investment guide or to take the place of advice from a qualified financial planner or other professional. Given the risk involved in investing of almost any kind, there is no guarantee that the investment methods suggested in this book will be profitable. Thus, neither the publisher nor the author assume liability of any kind for any losses that may be sustained as a result of applying the methods suggested in this book, and any such liability is hereby expressly disclaimed.

HarperCollins books may be purchased for educational, business, or sales promotional use. For information, please email the Special Markets Department at SPsales@harpercollins.com.

FIRST EDITION

Designed by Leah Carlson-Stanisic

Art on title page courtesy of POJ STUDIO 9899/Shutterstock, Inc.

Library of Congress Cataloging-in-Publication Data has been applied for.

ISBN 978-0-06-265120-4

ISBN 978-0-06-269956-5 (International Edition)

17 18 19 20 21 LSC 10 9 8 7 6 5 4 3 2 1

TO MONEY

For the wonderful things you do *for* us, the terrible things you do *to* us, and all the gray matter in between

CONTENTS

PART III

Now What? Building on the Shoulders of Flawed Thinking

INTRODUCTION

In 1975, Bob Eubanks hosted a short-lived TV game show called *The Diamond Head Game.* Taped in Hawaii, it featured a unique bonus round called "The Money Volcano." Contestants were put in a glass box that quickly transformed into a furious wind tunnel of flying money. Bills whirled, spun, and flapped all around as the players scrambled to grab as much as they possibly could before time ran out. They went absolutely bonkers inside the Money Volcano, reaching, clutching, spinning, flailing about inside a tornado of cash. It was great entertainment: For fifteen seconds it was clear that nothing in the world was more important than money.

To a certain extent, we are all inside the Money Volcano. We are playing the game in a less intense and visible manner, but we have been playing, and being played, for many years, in countless ways. Most of us think about money a lot of the time: how much we have, how much we need, how to get more, how to keep what we have, and how much our neighbors, friends, and colleagues make, spend, and save. Luxuries, bills, opportunities, freedom, stress: Money touches every part of modern life, from family budgets to national politics, from shopping lists to savings accounts.

And there's more to think about every day, as the financial world becomes more advanced; as we get more complex mortgages, loans, and insurance; and as we live longer into retirement and face new financial technologies, more complex financial options, and greater financial challenges.

Thinking a lot about money would be fine if by thinking more about it we were able to make better decisions. But that's not the case. The truth is, making bad money decisions is a hallmark of humanity. We're fantastic at messing up our financial lives. Congratulations, humans. We're the best.

Consider these questions:

- Does it matter if we use credit cards or cash? We spend the same amount either way, right? Actually, studies show we are more willing to pay more when we use a credit card. We make bigger purchases and leave larger tips with credit cards. We're also more likely to underestimate or forget how much we spend when—you guessed it—using the payment method we use most: a credit card.
- What's a better deal, a locksmith who opens a door in two minutes and charges $100 or one who takes an hour and charges the same $100? Most people think the one who took longer is the better deal, because he put in more effort and he cost less per hour. But what if the locksmith who took longer had to try several times and broke a bunch of tools before he succeeded? And charged $120? Surprisingly, most people still think this locksmith is a better value than the speedy one, even though all he did was waste an hour of our time with his incompetence.
- Are we saving enough for retirement? Do we all know even vaguely when we'll stop working, how much we'll have earned and saved by then, how our investments will have grown and

what our expenses will be for the exact number of years we'll live after that? No? We're so intimidated by retirement planning that, as a society, we're saving less than 10 percent of what we need, aren't confident we are saving enough, and believe we'll have to work until we're eighty even though our life expectancy is seventy-eight. Well, that's one way to cut down on retirement expenses: Never retire.

- Do we spend our time wisely? Or do we spend more time driving around looking for a gas station that will save us a few cents than we spend trying to find a cheaper mortgage?

Not only does thinking about money not improve financial decision-making, but sometimes the simple act of thinking about money actually changes us in deep and troublesome ways.[1] Money is the top reason for divorce[2] and the number one cause of stress in Americans.[3] People are demonstrably worse at all kinds of problem solving when they have money problems on their mind.[4] One set of studies showed that the wealthy, particularly when reminded they are wealthy, often act less ethically than the average person,[5] while another study found that just seeing images of money makes people more likely to steal from the office, hire a shady colleague, or lie to get more money.[6] Thinking about money literally messes with our heads.

Given the importance of money—for our own lives, for the economy, and for society—and given the challenges we have thinking about money in rational ways, what can we do to sharpen the way we think? The standard answer to this question is usually "financial education" or the more sophisticated term, "financial literacy." Unfortunately, financial literacy lessons, like how to buy a car or get a mortgage, tend to fade quickly, with almost zero long-term impact on our actions.

So, this book is not going to "financially literate" us or tell us what to do with our money every time we open our wallets. Instead, we'll explore some of the most common mistakes we make when it comes to money, and, more important, why we make these mistakes. Then, when we face our next financial decision, we might be better able to understand the forces at play and, hopefully, make better choices. Or at least more informed ones.

We're going to introduce a bunch of people and share their money stories. We'll show what they did in certain financial situations. Then we'll explain what science tells us about their experiences. Some of these stories are real, while some are, like the movies, "based upon a true story." Some of the people are reasonable. Some are fools. They might seem to fit certain stereotypes because we'll emphasize, even exaggerate, some of their characteristics in order to highlight certain common behaviors. We hope everyone recognizes the humanity, the mistakes, and the promise in each of their stories and how they echo in our own lives.

This book reveals how we think about money and the mistakes we make when we do. It's about the gaps between our conscious understanding of how money works, the way we actually use money, and how we should rationally think about and use money. It's about the challenges we all have reasoning about money, and the common mistakes we make spending it.

Will we be able to spend our money more wisely after reading this book? For sure. Maybe. A little bit. Probably.

At a minimum, we believe that revealing the complex forces behind the money choices that consume our time and control our lives can improve our financial affairs. We also believe that by understanding money's impact on our thinking, we will be able to make better *nonfinancial* decisions. Why? Because our decisions about money are about more than just money. The same forces that shape our reality in the domain of money also influence how we value the important things in the rest of our lives: how we spend our time, manage our career, embrace other people, develop relationships, make ourselves happy, and, ultimately, how we understand the world around us.

Put more simply, this book is going to make everything better. Isn't that worth the cover price?

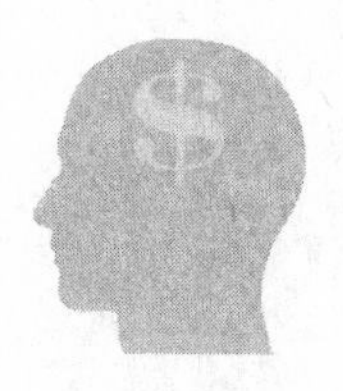

PART I

WHAT IS MONEY?

1

DON'T BET ON IT

George Jones* needs to blow off some steam. Work is stressful, the kids are fighting, and money is tight. So on a company trip to Las Vegas he heads to a casino. He parks, for free, in the lot at the end of a remarkably well-kept, publicly financed road and wanders aimlessly, head down, into the alternate universe of the casino.

The sound wakes him from his stupor: eighties music and cash registers mixed with clinking coins and the dinging of a thousand slot machines. He wonders how long he's been at the casino. There are no clocks, but judging by the old people slumped at the slot machines, it might have been a lifetime. It was probably five minutes. He couldn't be far from the entrance. But, then again, he can't see the entrance . . . or the exit . . . or any doors or windows or hallways or means of escape whatsoever. Just flashing lights, scantily clad cocktail servers, dollar signs, and people who are either ecstatic or miserable . . . but never anything in between.

Slot machines? Sure, why not? His first spin *just* misses a big score.

* Not the singer, but someone we made up. For our purposes, let's assume he can't sing at all. Not even karaoke.

So he spends fifteen minutes pumping in dollar bills to catch up. He never wins, but he does *just* miss quite a few more times.

Once his wallet is emptied of those pesky small-denomination bills, George grabs two hundred bucks at the ATM—not worrying about the $3.50 service fee because he'll cover that with his first winning hand—and sits down at a blackjack table. In exchange for ten crisp $20 bills, the dealer gives him a colorful pile of red plastic chips. There's a picture of the casino on them, with some feathers and an arrow and a teepee. They say $5, but they certainly don't feel like money. They feel like toys. George twirls them in his fingers, bounces them off the table, watches everyone's piles fluctuate, and covets the dealer's rainbow stash. George asks her to be kind to him. "Honey, as far as I'm concerned, you can have all of it—it ain't mine."

A cute, friendly server brings George a free drink. Free! What a deal! He's already winning. He tips her one little plastic toy chip.

George plays. George has some fun. George has some of the opposite of fun. He wins a little, loses more. Sometimes, when the odds seem to be in his favor, he doubles down or splits his cards, risking four chips instead of two, six instead of three. He ends up losing his $200. Somehow he avoids duplicating his tablemates' feats of amassing giant stacks of chips one minute, then unfurling reams of bills to buy more the next. Some of them are good-natured, some get angry when others "take their card," but none seem like the type who can afford to lose $500 or $1,000 in an hour. Still, this happens time and time again.

Earlier that morning, George had turned around just ten steps from his local café because he could save $4 by brewing coffee back at his hotel room. This evening, he tossed away forty $5 chips without blinking. Heck, he even gave the dealer one for being so nice.

What's Going On Here?

Casinos have perfected the art of separating us from our money, so it's a little unfair of us to start here. Nonetheless, George's experience gives us a quick glimpse into some of the psychological mistakes we make, even in less malicious settings.

The following are a few of the factors at play under the dazzling lights of the casino floor. We'll get into each of these in much more detail in the chapters to come:

Mental Accounting. George is worried about his finances—as evidenced by his decision to save money on coffee in the morning—yet nonchalantly spends $200 at the casino. This contradiction occurs, in part, because he puts that casino spending into a different "mental account" than the coffee. By taking his money and converting it into pieces of plastic, he opens an "entertainment" fund, while his other spending still comes out of something like "daily expenses." This trick helps him to feel differently about the two types of spending, but they're all really part of one account: "George's money."

The Price of Free. George is excited to get free parking and free drinks. Sure, he's not paying for them directly, but these "free" things get George to the casino in a good mood and impair his judgment. These "free" items, in fact, extract a high cost. There is a saying that the best things in life are free. Maybe. But free often ends up costing us in unexpected ways.

The Pain of Paying. George doesn't feel like he's spending money when he uses the colorful casino chips to gamble or tip. He feels like he's playing a game. Without feeling the loss of money with every chip, without being fully aware that he's spending it, he becomes less conscious of his choices and less considerate of the implications of his decisions. Spending plastic doesn't feel real the way that handing over paper bills would, so he keeps tossing them away.

Relativity. That $5 tip George gave the server—on a free drink—and his $3.50 ATM fee don't seem consequential compared to the stacks of chips surrounding him at the blackjack table or the $200 he was simultaneously taking out at the ATM. Those are *relatively* small amounts of money, and because he is thinking about them in relative terms, it is easier for him to go ahead and spend. Earlier in the day, on the other hand, the $4 coffee, compared to the $0 coffee at his hotel room, felt relatively too much to spend.

Expectations. Surrounded by the sights and sounds of money—cash registers, bright lights, dollar signs—George fancies himself a James Bond, 007, inevitable, suave victor over long casino odds and super-villains alike.

Self-Control. Gambling, of course, is a serious issue—an addiction, even—for many people. For our purposes, however, we can simply say that George, influenced by his stress and surroundings, the friendly staff, and "easy" opportunities, has a hard time resisting the immediate temptations of gambling for the distant benefits of having $200 more when he retires.

All of these mistakes may seem like they're unique to a casino, but in truth, the whole world is a lot more like a casino than we'd like to admit: In 2016, America even elected a casino owner as president, after all. Although we don't all blow off steam by gambling, we do all face similar decision-making challenges in terms of mental accounting, free, the pain of paying, relativity, self-control, and more. The mistakes George makes in the casino happen in many aspects of our daily lives. These mistakes are fundamentally rooted in our basic misunderstanding of the nature of money.

Although most of us probably believe we have a decent grasp of money as a topic, the surprising truth is, we really don't understand what it is and what it does for us, and, more surprisingly, what it does to us.

2

OPPORTUNITY KNOCKS

So, what exactly is money? What does it do for us and to us?

Those thoughts surely never crossed George's mind at the casino, and rarely, if ever, do they cross our minds. But they are important questions to ask and a great place to start.

Money represents *VALUE*. Money itself has no value. It only represents the value of other things that we can get with it. It's a messenger of worth.

That's great! Money makes it easy to value goods and services, which makes it easy to exchange them. Unlike our ancestors, we don't have to spend a lot of time bartering, plundering, or pillaging to get basic necessities. That's good, because few of us are handy with a crossbow or a catapult.

There are certain special features of money that make it extra useful:

- It is **general**: We can exchange it with almost everything
- It is **divisible**: It can be applied to almost any item of any size, no matter how large or small.

- It is **fungible**: We don't need a specific piece of currency, because it can be replaced by any other piece representing the same amount. Any $10 bill is as good as any other $10 bill, no matter where and how we get it.
- It is **storable**: It can be used at any time, now or in the future. Money doesn't age or rot, unlike cars, furniture, organic produce, or college T-shirts.

In other words, any amount of any money can be used at any time to buy (almost) anything. This essential fact helped us humans—*Homo irrationalis*—to stop bartering with each other directly and, instead, use a symbol—money—to exchange goods and services with much greater efficiency. That, in turn, gives money its final and most important feature: It is a *COMMON GOOD*, which means it can be used by anyone and for (almost) anything.

When we consider all of these characteristics, it is easy to see that there would be no modern life as we know it without money. Money allows us to save, to try new things, to share, and to specialize—to become teachers and artists, lawyers and farmers. Money frees us to use our time and effort to pursue all kinds of activities, to explore our talents and passions, to learn new things, and to enjoy art and wine and music, which themselves would not exist to any great extent without money.

Money has changed the human condition as much as any other advance—as much as the printing press, the wheel, electricity, or even reality television.

While it is important to recognize how important and useful money is, unfortunately some of money's benefits are also the source of its curses. They create many of the difficulties that come with it. As the great philosopher Notorious B.I.G. said, "Mo' Money Mo' Problems."

To consider the blessings and curses of money—that indeed there are two sides to every coin, pun intended—let's think about the general nature of money. There is no question that the ability to exchange money with an almost infinite variety of things is a crucial and wonderful thing, but it also means that the complexity of making decisions about money is incredibly high.

Despite the popular expression, comparing apples to oranges is actually quite easy. If we're standing next to a fruit plate with an orange and an apple, we know exactly which one we want at any particular moment. If money is involved, however, and we must decide if we're willing to pay $1 or 50 cents for that apple, it is a harder decision. If the price of the apple is $1 but the orange costs 75 cents, the decision gets even more complex. Whenever money is added to *any* decision, it gets more complex!

Opportunity Lost

Why do these money decisions become more complicated? Because of *OPPORTUNITY COSTS.*

When we take the special features of money into account—that money is general, divisible, storable, fungible, and, especially, that it is the common good—it becomes clear that we really can do almost anything with money. But just because we can do almost *any*thing with it, that doesn't mean we can do *every*thing. We must make choices. We must make sacrifices; we must choose things *not* to do. That means, we absolutely must, consciously or not, consider opportunity costs every time we use money.

Opportunity costs are alternatives. They are the things that we give away, now or later, in order to do something. These are the opportunities that we sacrifice when we make a choice.

The way we *should* think about the opportunity cost of money is that when we spend money on one thing, it's money that we cannot spend on something else, neither right now nor anytime later.

Imagine, once again, that we're in front of that fruit plate, but now we're in a world that has only two products—an apple and an orange. The opportunity cost of buying an apple is a forgone orange, and the opportunity cost of buying an orange is the forgone apple.

Similarly, the $4 our casino friend George might have spent at his local café could be bus fare, or part of lunch, or snacks at the Gamblers Anonymous meetings he'll attend in a few years. He wouldn't have been giving up $4; he would have given up opportunities that those dollars could have provided either now or in the future.

To get a better idea of both the importance of opportunity cost and why we fail to take it sufficiently into account, pretend you're given $500 each Monday and that that is all the money you can spend that week. In the beginning of the week, you may not consider the consequences of your decisions. You don't realize what you are giving up when you buy dinner and have a drink or buy that beautiful shirt you've had your eye on. But as the $500 dwindles and Friday rolls around, you find yourself with only $43 left. Then it becomes much clearer that opportunity costs exist and that what you spent early in the week is now affecting what you have left to spend. Your decision to pay for dinner, drinks, and the snazzy shirt on Monday leaves you with a tough choice on Sunday—you can afford to either buy the newspaper or eat a bagel with cream cheese, but not both. On Monday, you had an opportunity cost to consider, but it wasn't as clear to you. Now, on Sunday, when the opportunity cost is finally clear, it is too late (though, on the bright side, at least you probably look good reading the sports section on an empty stomach).

So, opportunity costs are what we *should* think about as we make financial decisions. We should consider the alternatives we are giving up by choosing to spend money now. But we don't think about opportunity costs enough, or even at all. That's our biggest money mistake and the reason we make many other mistakes. It is the shaky foundation upon which our financial houses are built.

A BIGGER PICTURE

Opportunity costs are not restricted to the realm of personal finance. They have global ramifications, as President Dwight Eisenhower noted in a 1953 speech about the arms race:

> **Every gun that is made, every warship launched, every rocket fired signifies, in the final sense, a theft from those who hunger and are not fed, those who are cold and are not clothed. This world in arms is not spending money alone. It is spending the sweat of its laborers, the genius of its scientists, the hopes of its children. The cost of one modern heavy bomber is this: a modern brick school in more than thirty cities. It is two electric power plants, each serving a town of sixty thousand population. It is two fine, fully equipped hospitals. It is some fifty miles of concrete pavement. We pay for a single fighter plane with a half million bushels of wheat. We pay for a single destroyer with new homes that could have housed more than eight thousand people.**

Thankfully, most of our personal dealings with opportunity costs lie closer to the price of an apple than the cost of war.

A few years ago, Dan and a research assistant went to a Toyota dealership and asked people what they would give up if they purchased a new car. Almost no one had an answer. None of the shoppers had spent any significant time considering that the thousands of dollars they were about to spend on a car could be spent on other things. So, Dan tried to push a little bit further with the next question, and asked what specific products and services they wouldn't be able to get if they went ahead and bought that Toyota. Most people answered that if they bought a Toyota, they wouldn't be able to buy a Honda, or some other simple substitution. Few people answered that they wouldn't be able to go to

Spain that summer and Hawaii the year after, or that they wouldn't go out to a nice restaurant twice a month for the next few years, or that they would be paying their college loans for five more years. They were seemingly unable or unwilling to think of the money they were about to spend as their potential ability to buy a sequence of experiences and goods over time in the future. This is because money is so abstract and general that we have a hard time imagining opportunity costs or taking them into account. Basically, nothing specific comes to mind when we spend money except the thing we're contemplating buying.

Our inability to consider opportunity costs, as well as our general resistance to considering them, is not limited to car shopping. We almost always fail to fully appreciate alternatives. And, unfortunately, when we fail to consider these opportunity costs, the odds are that our decisions are not going to be in our best interests.

Consider the experience of buying a stereo system, as conveyed by Shane Frederick, Nathan Novemsky, Jing Wang, Ravi Dhar, and Stephen Nowlis in an aptly named paper, "Opportunity Cost Neglect." In their experiment, one group of participants was asked to decide between a $1,000 Pioneer and a $700 Sony. A second group was asked to pick between the $1,000 Pioneer and a package deal where for $1,000 they could get the Sony *plus* $300 to be spent only on CDs.

In reality both groups were choosing between different ways of spending that $1,000. The first group chose between spending all of it on a Pioneer or spending $700 on a Sony and $300 on other things. The second group chose between spending all of it on a Pioneer or spending $700 on a Sony and $300 on music. The results showed that the Sony stereo was a much more popular choice when it was accompanied by $300 of CDs than when it was sold without them. Why is this odd? Well, strictly speaking, an unconstrained $300 is worth more than $300 that must be spent on CDs because we can buy anything with the unconstrained money—including CDs. But when the $300 was framed as being dedicated to CDs, the participants found it more appealing. That's because $300 worth of CDs is much more concrete and defined than just $300 of "anything." In the $300-for-CD case we know what we're getting. It is tangible and easy to evaluate. When the $300 is abstract and general, we don't conjure up the specific images of how we're going to spend it, and the emotional, motivational forces on us are less powerful. This is just one more example of how when we represent money in a general way, we end up undervaluing it compared to when we have a specific representation of that money.[1]

Yes, CDs are the example here, which nowadays is like thinking about the gas efficiency of a stegosaurus, but the point remains: People are somewhat surprised when we simply remind them that there are alternative ways to spend money, whether it's on a vacation or on a pile of CDs. That surprise suggests that people don't tend to naturally consider alternatives, and without considering alternatives, we can't possibly take opportunity costs into account.

This tendency for neglecting opportunity costs shows us the basic flaw in our thinking. It turns out that the wonderful thing about money—that we can exchange it for so many different things now and in the future—is also the biggest reason that our behavior around money is so problematic. While we should be thinking about spending in terms of opportunity cost—that spending money now on one thing is a trade-off for spending it on something else—thinking this way is too abstract. It's too hard. So we simply don't do it.

To make matters worse, modern life has given us endless financial instruments, such as credit cards, mortgages, car payments, and student loans, which further—and often purposefully—obscure our ability to understand the future effects of spending money.

When we cannot, or will not, think about money decisions the way we should, we fall back on all kinds of mental shortcuts. Many of these strategies help us deal with the complexity of money, though they don't necessarily help us do so in the most desirable or logical ways. And they often lead us to value things incorrectly.

3

A VALUE PROPOSITION

Jeff's young son recently asked him for a story while they were on a plane. The children's books were in the checked bags—even though his wife had explicitly said to put them in the carry-on! So Jeff made up the following derivative of Dr. Seuss's *There's a Wocket in My Pocket!*

> How much would you pay for a dribble? A zabble? A gnabble? A quibble?
>
> What about zork? A nork? An imported Albanian three-toed blork?

While it may seem like Jeff was just torturing nearby passengers (not to mention his kid), how different are those questions from those we face in real life?

How do we know what we'd pay for a "Coca-Cola," or a month of "Netflix," or an "iPhone"? What are these words? What are these things? How do we value items that, to a visitor from another planet, would seem as nonsensical as a Zamp behind a Lamp or a Yottle in a

Bottle? If we had no idea what something was, what the price was, or what other people had actually paid for it, how would we know what to pay for these things?

What about art? How is a Jackson Pollock painting any different from an imported Albanian three-toed blork? It's just as unique and unusual . . . and probably just as practical. Yet art somehow has a price. In 2015, a buyer spent $179 million on what the *New Yorker* called "a so-so Picasso, from his just-O.K. later period."[1] Another guy took people's Instagram pictures—posted online and viewable *for free*—blew them up, and sold them for $90,000.[2] There was even a photograph of a potato that sold for 1 million euros. Who sets these prices? How are these values determined? Would anyone like to buy a picture of some potatoes we just took with our phone?

We've all undoubtedly heard a lot about "value." Value reflects the worth of something, what we might be willing to pay for a product or service. In essence, value *should* mirror opportunity cost. It should accurately reflect what we're willing to give up in order to acquire an item or experience. And we *should* spend our money according to the actual value of different options.

In an ideal world, we'd accurately assess the value of every purchase. "What is this worth to me? What am I willing to give up for it? What is the opportunity cost here? That is what I will pay for it." But, as fitness magazines remind us, we don't live in an ideal world: We don't have six-pack abs and we don't accurately assess value.

Here are just a few of the historical ways in which humans have valued things incorrectly:

- The Native Americans sold Manhattan for some beads and guilders. How could they have known how to value something—property—that they had never heard of, and for which they had no context?
- The cost to rent an apartment in some major cities can climb to more than $4,000 per month, and we don't seem to blink. The price of gas rises 15 cents, and it can swing a national election.

- We pay $4 for a coffee at a "café" when the same basic drink is available for $1 in a convenience store next door.
- Start-up tech companies with no revenue are regularly valued to be worth hundreds of millions, even billions, of dollars, and we act surprised when they don't live up to these expectations.
- Some people go on a $10,000 vacation but spend twenty minutes each day looking for free parking.
- We comparison shop for smartphones. We think we have an idea of what we're doing, and at the end, we feel we have made the right choice.
- King Richard III was willing to sell his kingdom, *his entire kingdom*, for a horse. His kingdom for a horse!

We have always assessed value in ways that are not necessarily connected to value at all.

If we were perfectly rational creatures, a book about money would be about the value we place on products and services because, rationally, money equals opportunity costs equals value. But we are not rational, as noted in Dan's other books (*Predictably Irrational, The Upside of Irrationality, Hey Guys! We Are Sooooo Not Rational!*[*]). Rather, we use all kinds of quirky mental tricks to figure out how much we value things—that is, how much we are willing to pay. Thus, *this* book is about the odd, wild, and, yes, completely irrational ways we approach spending decisions and about the forces that cause us to overvalue some things and undervalue others.

We think of these forces, these tricks and shortcuts, as "value cues." They are cues that we believe are associated with the real value of a product or service but often are not. Sure, some value cues are fairly accurate. But many are irrelevant and misleading and others are intentionally manipulative. And yet, we allow these cues to change our perception of value.

[*] Not a real title. Yet.

Why? It's not because we like making mistakes or inflicting pain on ourselves (although there are places where we can pay for that, too). We follow these cues because it is so hard to consider opportunity costs and assess real value. Moreover, it becomes ever harder to figure out how much we are willing to pay for something when the financial world is trying to confuse and distract us.

This dynamic is key: We are, of course, constantly fighting the complex nature of money and our own failure to consider opportunity costs. Worse, we are also constantly fighting external forces trying to get us to spend more, more frequently, and more freely. There are numerous forces that want us to incorrectly assess true value, because it profits them when we spend irrationally. Given all the challenges we face, it's a wonder we're not all wandering around billion-dollar studio apartments drinking Yottle in a Bottle from a thousand-dollar Blork.

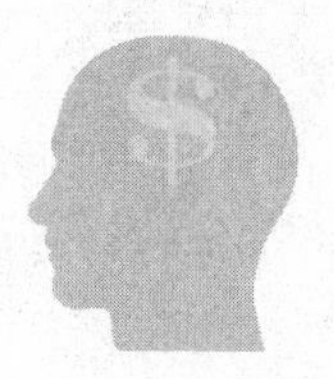

PART II

HOW WE ASSESS VALUE IN WAYS THAT HAVE LITTLE TO DO WITH VALUE

4

WE FORGET THAT EVERYTHING IS RELATIVE

Susan Thompkins is somebody's Aunt Susan, and everyone has a version of someone like Aunt Susan. Aunt Susan is a genuinely happy and loving woman, who also buys gifts for her nephews and nieces whenever she shops for herself and her kids. Aunt Susan loves shopping at JCPenney. She's been shopping there since she was a child, going with her parents and grandparents, helping them spot bargains. There were always so many great deals to be found. It was a fun game, running around, looking for the highest number next to the percent symbol, proud of spotting the secret stash.

In recent years, Aunt Susan would drag along her brother's kids, showing them ugly sweaters and mismatched outfits that they just "couldn't pass up because they're such great deals!" While the kids didn't love it, she did. Getting the great bargains at JCPenney was still a big thrill for Aunt Susan.

Then, one day, Ron Johnson, JCPenney's new CEO, got rid of all of the deals. He instituted what he called "fair and square" pricing across the board. No more sales, bargains, coupons, or discounts.

Suddenly Susan was sad. Then she was angry. Then she stopped going to JCPenney entirely. She even formed an online group with her friends called "I hate Ron Johnson." She wasn't alone. Many customers left JCPenney. It was a bad time for the company. It was a bad time for Susan. It was a bad time for Ron Johnson. It was a bad time for the ugly sweaters, too: They couldn't buy themselves. The only ones having a good time? Susan's nephews.

A year later, Aunt Susan heard discounts had returned to JCPenney. Cautiously, with her guard up, she returned. She hunted through a rack of pantsuits, examined some scarves, and checked out a paperweight display. And she looked at the prices. "20% off." "Marked down." "For sale." She bought just a couple of things that first day, but since then, she's returned to her old JCPenney self. She's happy again. And that means more shopping trips, ugly sweaters, and awkward thank-yous from her loved ones. Hooray.

A JCPenney for Your Thoughts

In 2012, Ron Johnson, the new CEO of JCPenney, did scrap Penney's traditional, and yes, slightly deceptive practice of marking products up and then marking them back down. In the decades before Johnson's arrival, JCPenney always offered customers like Aunt Susan coupons, deals, and in-store discounts. These reduced Penney's "regular prices," which were artificially inflated, to appear to be "bargain deals," but in fact, after the discounts, their prices were in line with prices everywhere else. In order to get to the final, retail price of an item, customers and the store would perform this Kabuki theater of raising prices at first and then lowering them in all kinds of creative ways, with different signs and percentages and sales and discounts. And they played this game over and over again.

Then Ron Johnson made the store's prices "fair and square." No more coupon cutting, bargain hunting, and sale gimmicks. Just the real price, roughly equal to those of its rivals and roughly equal to their previous "final" prices—after raising and discounting them. Johnson believed his new practice was clearer, more respectful, and less manipulative for his customers (and he was right, of course).

Except that loyal customers like Aunt Susan hated it. They detested "fair and square." They abandoned the chain, grumbling about feeling cheated, being misled and betrayed by the real and true cost, and not liking the honest, fair-and-square pricing. Within a year, JCPenney lost an amazing $985 million and Johnson was out of a job.

Almost immediately after his firing, the list price of most items at JCPenney rose by 60 percent or more. One side table that cost $150 rose to an "everyday price" of $245.[1] Not only were the regular prices higher, but there were more discount options: Instead of just a single dollar amount, the store offered "sale," "original," and "appraised at" prices. Of course, when we factor in the discounts available—by sale, or coupon, or special deal—the prices pretty much stayed the same. They just didn't look that way. Now it looked like JCPenney was once again offering really great deals.

Ron Johnson's JCPenney offered products at more honest prices and was rejected in favor of sales gimmicks. Aunt Susan still hates him. Think about that: JCPenney's customers voted with their wallets and they elected to be manipulated. They wanted deals, bargains, and sales, even if it meant bringing back inflated regular prices—which is exactly what JCPenney eventually did.

JCPenney—and Ron Johnson—paid a high price for failing to understand the psychology of pricing.* But the company ultimately learned that it could build a business based upon our inability to assess

*If you happen to run a large chain of department stores and ever contemplate making wholesale, fundamental changes to your pricing, we humbly suggest you test it at a single store or two before implementing it everywhere. Unless you are looking to get fired so you can claim a nice severance package, in which case, we abstain from offering advice.

value rationally. Or, as H. L. Mencken once said, "No one ever went broke underestimating the intelligence of the American public."

What's Going On Here?

The story of Aunt Susan and JCPenney shows some of the many effects of *RELATIVITY*, one of the most powerful forces that make us assess value in ways that have little to do with actual value. At JCPenney, Aunt Susan assessed value based upon *relative* value, but relative to what? Relative to the original posted price. JCPenney helped her make the comparison by posting the discount as a percentage and adding notes like "sale" and "special" to help focus her attention on the amazing relative price they offered.

Which would you buy? A dress shirt priced at $60 or the very same dress shirt, priced at $100, but "On Sale! 40% off! Only $60!"?

It shouldn't matter, right? A $60 shirt is a $60 shirt, no matter what language and graphics are on the price tag. Yes, but since relativity works on us at a very deep level, we don't see these two in the same way, and if we were a regular like Aunt Susan, we would buy the on-sale shirt every time—and be outraged by the mere presence of the straight-up $60 one.

Is this behavior logical? *No.* Does it make sense once you understand relativity? *Yes.* Does it happen frequently? *Yes.* Did it cost an executive his job? *Absolutely.*

We often cannot measure the value of goods and services on their own. In a vacuum, how could we figure the cost of a house or a sandwich, medical care or an Albanian three-toed blork? The difficulty of figuring out how to value things correctly makes us seek alternative ways to measure value. That's where relativity comes in.

When it is hard to measure directly the value of something, we compare it to other things, like a competing product or other versions of the same product. When we compare items, we create relative values. That doesn't seem too problematic, right?

The problem isn't with the concept of relativity itself, but with the way we apply it. If we compared everything to all other things, we would consider our opportunity costs and all would be well. But we don't. We compare the item to only one other (sometimes two). This is when relativity can fool us.

Sixty dollars is relatively cheap compared to $100, but remember opportunity costs? We should be comparing $60 to $0, or to all of the other things we could buy with $60. But we don't. Not when, like Aunt Susan, we use relative value to compare the current price of an item to the amount it used to cost before the sale (or was said to cost) as a way to determine its value. This is how relativity confounds us.

JCPenney's sale prices offered an important value cue to customers. Not just an *important* cue, but often the *only* cue. The sale price—and the savings JCPenney touted—provided customers context for how good a deal each purchase was.

JCPenney's sale signs provided customers with context, and without context, how could we determine the value of a shirt? How could we know whether it's worth $60 or not? We can't. But compared to a $100 shirt, a $60 one sure seems like a great value, doesn't it? Why, it's almost like getting $40 for free! Let's all buy one so our nephews can be mocked at school!

By eliminating the sales and "savings," JCPenney removed an element that helped their customers feel that their decisions were the right ones. Just looking at a sale price next to a "regular" price gave them some indication that they were making a smart decision. But they weren't.

Relatively Speaking

Let's step away from our wallets for a second and consider the principle of relativity more generally.

One of our favorite optical illusions is this image of black and gray circles:

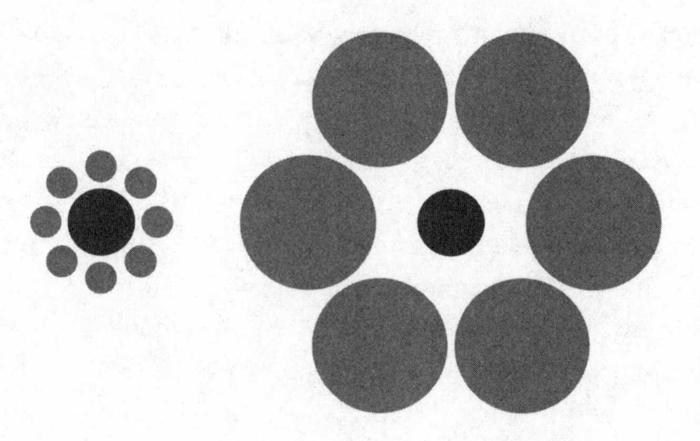

It's pretty obvious that the black circle on the right is smaller than the one on the left, right? The thing is, it's not. Both black circles are exactly, and almost unbelievably, the same size. Go ahead, disbelievers: Cover up the gray circles and compare. We'll wait.

The reason this illusion fools us is that we don't compare the two black circles directly to each other, but rather to their immediate surroundings. In this case, that's the gray circles. The black on the left is large compared to its gray circles and the black on the right is small compared to its circles. Once we've framed their sizes this way, the comparison between the two black circles is between their relative, rather than absolute, size. That's visual relativity.

And because we love visual illusions so much, here is another one of our favorites, the Adelson checker illusion. It involves a basic checkerboard with a cylinder on one side casting a shadow over the squares. (In keeping with the theme of this chapter, our version uses an ugly sweater instead of a cylinder.) Two squares are labeled. Square A lies outside the shadow, while B is inside. When we compare them, it's quite clear that A is much darker, right? The thing is, it's not. A and B are exactly, and almost unbelievably, the same shade. Go ahead, disbelievers: Use something to cover all the other squares. Now compare A and B. We'll wait.

Relativity works as a general mechanism for the mind, in many ways and across many different areas of life. For example, Brian Wansink, author of *Mindless Eating*,[2] showed that relativity can also affect our waistlines. We decide how much to eat not simply as a function of how much food we actually consume, but by a comparison to its alternatives. Say we have to choose between three burgers on a menu, at 8, 10, and 12 ounces. We are likely to pick the 10-ounce burger and be perfectly satisfied at the end of the meal. But if our options are instead 10, 12, and 14 ounces, we are likely again to choose the middle one, and again feel equally happy and satisfied with the 12-ounce burger at the end of the meal, even though we ate more, which we did not need in order to get our daily nourishment or in order to feel full.

People also compare food to other objects in their environments. For instance, people compare the amount of food to the size of the plate. In one of Brian's experiments, he connected soup bowls to the table, asking people to eat soup until they had had enough. Some people simply ate soup until they did not want anymore. But one group of participants were unknowingly eating from bowls that had tiny hoses connected to the bottom. As they ate, Brian was slowly pushing a bit of soup into their bowls at an imperceptible rate. Every spoon of soup out, a bit of soup went in. In the end, those who got the endless soup bowls ate much more soup than those with normal, nonreplenishing bowls. And when he stopped them after they ate a lot of soup (and he had to stop them), they said that they were still hungry. The endless-soup-bowl recipients didn't get their cues for satisfaction from how much soup they'd consumed or how hungry they felt. Rather, they judged their satisfaction by the level of reduction they saw relative to the bowl. (Speaking of relatives, were we to conduct a similar experiment around family gatherings, many of us might keep eating just so we didn't have to talk to our cousins, uncles, aunts, parents, and grandparents. But that's a different kind of relativity.)

This kind of comparison isn't confined to objects in the same basic category, like soup or hamburgers, either. When Italian diamond dealer Salvador Assael first attempted to sell the now-popular Tahitian black pearls, not a single buyer bit. Assael did not give up, nor did he merely throw some black pearls in with shipments of white ones, hoping they might catch on. Instead, he convinced his friend, jeweler Harry Winston, to feature the black pearls in his Fifth Avenue store window surrounded by diamonds and other precious stones. In no time, the pearls were a hit. Their price skyrocketed. A year earlier, they were worth nothing—probably less than the oysters they came from. Suddenly, however, the world believed that if a black pearl is deemed classy enough to be exhibited next to an elegant sapphire pendant, it must be worth a lot.

These examples show that relativity is a basic computation of the human mind. If it affects our understanding of value of concrete things like food and luxury jewelry, it also probably informs the way we think about what to do with our money in very powerful ways.

Relatively Common Financial Relatives

Besides Aunt Susan's bargain obsession, let's think about a few of the many ways in which we might let relative value obscure real value.

- At a car dealership, we get offered add-on options like leather seats and sunroofs, tire insurance, silver-lined ashtrays, and the useless pitch of the stereotypical car salesmen: undercoating. Car dealers—perhaps the most devious group of amateur psychologists this side of mattress salesmen—know that when we're spending $25,000, additional purchases, like a $200 CD changer, seem cheap, even inconsequential, in comparison. Would we ever buy a $200 CD changer? Does anyone even listen to CDs anymore? No and no. But at just 0.8 percent of the total purchase price, we hardly shrug. Those hardly-shrugs can add up quickly.
- When vacationing at a posh resort, we often don't get upset when we're charged $4 for a soda, even though it costs $1 elsewhere. In part, this is because we're lazy and like to lounge around like beached royalty. But it's also because, compared to the thousands of dollars we're spending on the rest of our tropical getaway, $4 seems like relatively small change.
- Supermarket checkout lines dare us to resist trashy tabloids and sugary candy, using the same approach. Compared to $200 for a week of food, $2 for a box of Tic Tacs or $6 for a magazine of Kardashians seems to be no big deal.
- Don't forget the wine! Fine vino in restaurants costs a lot more than it does in a wine shop. It's logical to pay more for the convenience of wine with dinner—we don't want to take a bite, then have to run to our car to swig from our dime-store Beaujolais—but it's also a tribute to relative versus absolute value. We might not pay $80 for a midlevel bottle of wine when we're also buying nachos and a spray can of processed cheese, but if we're dining at the exclusive French Laundry, paying several hundred dollars for the food, $80 doesn't seem like that much more for a drink. If you do manage to get a reservation at the famous California restaurant, however, it would be best to invite the authors of this book to join the dinner, just to confirm this hypothesis.

Speaking of supermarkets, Jeff recently had an interesting experience while shopping. For years, his favorite cereal was Optimum Slim. For a man of soft, round middle, advancing years, and limited exercise ambition, it promised just the right amount of slim. The optimum amount.

It had always cost $3.99 at his local store. Then, one day, he looked in the usual spot and couldn't find it. He looked and looked. No dice. He had a mini panic attack—a frequent occurrence, brought on by everything from missing breakfast food to lost TV remotes—until a clerk pointed to a new box in the old spot. There was a cereal there with the name "Nature's Path Organic—Low Fat Vanilla" and in the upper left corner a tiny picture of the old Optimum Slim box and a caption, "New Look—Same Great Taste."

Phew. He put down the Valium and picked up a box. Then a sign on the shelf caught his eye. "Nature's Path Organic Optimum Slim—Regular $6.69. *SALE $3.99.*"

Yup, his favorite cereal, which had always cost $3.99, now had a new look *and* a new price of . . . $3.99. Down from its "regular" price of . . . $6.69? It's one thing if the company introduced new packaging as a reason to raise the price. It's another thing if the store pretended the regular price was a sale in order to boost orders. But to do both at the same time—that's using a certain amount of relativity. The optimum amount.

The store and cereal company weren't trying to entice Jeff with this sign. He already liked the cereal. They were after new customers who had no way to judge the value of this "new" cereal. Without any context—without a way to know if it's tasty or healthy or what it's worth—they hoped customers would be impressed by the new name and make the easy comparison between $6.69 and $3.99 and decide, "Wow, this cereal, right now, has great value!"

Say we encounter something we've always wanted. Let's call it a **widget** (a common term in traditional economics textbooks representing a generic product designed both to obscure the fact that it has questionable value and to torment readers of traditional economics textbooks). Our widget is on sale! Fifty percent off! Exciting, right? But stop for a second. Why do we care about the sale? Why do we care about what it *used* to cost? It shouldn't matter what the cost was in the past since that's not what it costs now. But because we have no way of really knowing how much this precious widget is worth, we compare the price *now* to the price before the sale (called the "regular" price), and take that as an indicator of its high current amazing value.

Bargains also make us feel special and smart. They make us believe we're finding value where others haven't. To Aunt Susan, saving $40 on a $100 shirt seemed like getting $40 to spend elsewhere. On a more rational level, we shouldn't measure the value of what we are *not* spending—the $40—but rather the $60 we *are*. But that's not how we operate and that's not what we do.

Another place we see this kind of comparison is with quantity (so-called bulk) discounts. If a bottle of expensive shampoo is $16 and one twice the size is $25, all of a sudden the larger, more expensive bottle looks like a great deal, making it easy to forget the question of whether we really need that much, or that brand of, shampoo in the first place. Moreover, the bulk discounting practice also serves to hide the fact that we have no clue how to value the cocktail of chemicals that make up shampoo.

Had Albert Einstein been an economist rather than a physicist, he might have changed his famous theory of relativity from $E = MC^2$ to *$100 > Half Off of $200.*

Dollars and Percents

We might look at those examples and think, "Okay, I understand how using relativity is a mistake." That's good! "Buuuuuuut . . ." you're probably saying, "Those choices make sense because, as a percentage of what I'm spending, the extra expenditures are tiny." Well, yes, but a dollar should be a dollar, no matter what else we're spending or doing. Spending $200 on a CD player just because we happen to be buying a $25,000 car is the same irrelevant reasoning as spending $200 on a CD player just because we happen to be wearing a plaid shirt. It just doesn't *feel* as irrelevant.

Imagine we set out one Saturday morning with two errands. First, we're going to buy the running shoes we've been eyeing for a while. We go to the store and pick up the $60 sneakers. The person helping us confides that at another store down the street the same exact pair is on sale for $40. Is it worth driving five minutes to save $20? If we're like most people, the answer is yes.

Now that we've got our shoes, we embark on our second errand. We're going to buy patio furniture because it's finally spring! We find the perfect set of chairs and an umbrella-topped table at the garden store for $1,060. Once again, an employee tells us of a sale at another location that's five minutes away. We can get the same set there for $1,040. Do we spend five minutes to save $20 this time? If we're like most people, the answer, this time, is no.

In both cases, we don't look at the true, absolute value presented to us: $20 for a five-minute drive. Instead, we consider $20 compared to $60 and to $1,060 respectively. We compare the relative advantage of $40 shoes to $60 shoes, and decide the money is worth the time. Then we compare the relative advantage of a $1,040 patio set to a $1,060 one and find it's not. The first is a 33 percent savings, the second is 1.9 percent—yet the $20 of money saved is, in each case, identical.

This is also why the shopper who didn't shrug at the $200 CD changer on a $25,000 car might clip coupons to save 25 cents on a bag of chips or debate about a dollar or two tip at a restaurant. When relativity comes into play, we can find ourselves making quick decisions about large purchases and slow decisions about small ones, all because we think about the percentage of total spending, not the actual amount.

Are these logical choices? *No.* Are they the right choices? *Often not.*

Are they the *easy* choices? *Absolutely.* Most of us take the easy choice, most of the time. That's one of our big problems.

Easy Does It

Which question would we answer more quickly and decisively: "What do you want for dinner?" or "Do you want chicken or pizza for dinner?"

In the first, we're given endless options. In the second, we need only compare the two choices and decide which is relatively more appealing to us right now. The second question would get a quick response. It's an easier comparison. It's a trivial question, after all: Unless we're lactose intolerant, what kind of a monster chooses chicken over pizza? That's just crazy.

Relativity is built on two sets of decision shortcuts. First, when we can't assess absolute value, we use comparisons. Second, we tend to choose the *easy* comparison. Aylin Aydinli, Marco Bertini, and Anja Lambrecht studied relativity by looking at email sales such as Groupon offers—what they called "price promotions"—and found that they create a particularly telling emotional impact. Specifically, when we encounter price promotions, we spend less time considering different options. Furthermore, if we are later asked to recall details of the offer, we recall less product information.[3]

It seems that discounts are a potion for stupidity. They simply dumb down our decision-making process. When an item is "on sale," we act more quickly and with even less thought than if the product costs the same but is marked at a regular price.

Basically, since it is so hard for us to assess the real value of almost anything, when something is on sale—when we are presented with a relative valuation—we take the easy way out and make our decision based upon that sale price. Just as JCPenney customers loved to do, rather than trying to work hard and figure out an item's absolute value, when given the choice, we take the path of relatively least resistance.

DISTRACT AND DECOY

Relativity and our inclination to make the easy choice leave us susceptible to multiple types of external interventions and manipulations by those who set prices, including decoys. In *Predictably Irrational*, Dan used subscription offers to the *Economist* to illustrate the relativity problem. In that example, readers could get an online subscription for $59, a print subscription for $125, or a print *and* online subscription for $125.

If we're a smarty-pants, like the Massachusetts Institute of Technology graduate students Dan tested, 84 percent of us would choose the print *and* Web version for $125. None would choose the $125 print-only choice and only 16 percent would choose Web-only. Well, don't we look very smart in those pants?

But what if our choice was just between the $59 Web-only offer and the $125 print-and-Web option? Suddenly, if we were like those who paid thousands in tuition for a few extra years of doing problem sets at MIT, we'd act quite differently: 68 percent would choose Web only, while only 32 percent would go for the $125 print and Web, down from 84 percent in the first scenario.

Just by including the clearly inferior print-only option—*which no one chose*—the *Economist* nearly tripled sales of its $125 Web-and-print version. Why? Because that print-only option was a decoy employing relativity to push us toward the combo deal.

One hundred twenty-five dollars for print *and* Web is obviously a better choice than $125 for just print. We see that these two options are similar and easy to compare. They create relative value. We make our decision based on that comparison and feel smart about our choice. We feel even smarter once we read a few issues (and, sure, we'll look smarter to our friends when we leave a copy around the apartment). But how do we know we're not actually unwitting participants in a study proving that we're not so smart after all?

Dan's experiment showed how relativity can be (and often is) used against us. We compare print only to the print-and-Web combo because it is the simplest, most obvious, and easiest one to make. Because

SUBSCRIPTIONS

Welcome to
The Economist Subscription Centre

Pick the type of subscription you want to buy or renew.

❑ **Economist.com subscription** - US $59.00
One-year subscription to Economist.com
Includes online access to all articles from *The Economist* since 1997.

❑ **Print subscription** - US $125.00
One-year subscription to the print edition of *The Economist*.

❑ **Print & web subscription** - US $125.00
One-year subscription to the print edition of *The Economist* and online access to all articles from *The Economist* since 1997.

those options were most similar to each other in substance and price, they were simple to compare. That made it easy to forget, ignore, or avoid the other option, the one that would have required a more complex comparison. When we face easy comparisons we forget about the greater context, the alternative options—in this experiment, both the $59 option and the option of spending no money at all on the *Economist*. We follow the relativity path. We like to tell ourselves stories about why we do the things we do, and when we face relativity the story is easy to tell. We get sucked into justifying our actions this way, even when the justification makes little sense.

Another situation in which we find ourselves falling for the easy comparison—using relativity to assess value when there is no other simple way to do so—is when we have many choices and we can't easily evaluate any of them. Dan used the example of televisions: a 36-inch Panasonic for $690, a 42-inch Toshiba for $850, and a 50-inch Philips for $1,480. Faced with these choices, most people choose the middle option, the $850 Toshiba. The cheapest and most expensive items are road signs funneling us to the middle option. In this case, relativity doesn't compel us to compare one specific product to another; rather, it directs us toward specific product attributes, such as price or size, and gets us to look at the range of these attributes in a relative way. We say to ourselves: "The price ranges from $690 to $1,480" or "The size is between 36 and 50 inches." Then we pick relative to the range—often something in the middle.

When we have no idea what something should cost, we believe we're making the best decision if we neither overspend on the deluxe model nor go too cheap on the basic one. So we opt for the middle one, which is often what the marketers who set up the options wanted to sell us from the get-go. Even though we have no idea if that's what we wanted or if it's worth it, picking the middle choice just seems reasonable. It's not necessarily the wrong choice, but it is a choice made for reasons that have little to do with true value. It's like buying a $60 shirt because it used to cost $100, choosing the middle-sized burger whether the options are 8, 10, and 12 ounces or 10, 12, and 14 ounces, or buying a tub of popcorn for $8 at the movie theater just because they are also selling a $9 supertub that seems way too big. When there are two options, relativity is perfectly fine. Those decisions aren't about the absolute value of our choice, but about the relative alternatives.

So we often go for the easy comparison. Marketers, menu designers, and politicians know this, and use this trick when planning their strategies. Now we know this trick, too, and with this knowledge we can look at the world slightly more objectively. Now that you know, maybe the commercial playing field is slightly more leveled.

A Bundle of Oy

Relativity also affects our value assessments when products are bundled, that is, when products offer multiple features and options. In these situations, relativity seems to offer an escape from complexity. However, it actually creates the opportunity for another type of problem and more confusion.

Consider fast-food "value meals." We could order two separate items—but why not get them together and throw in a third for just a few pennies more? Want a hamburger and a soda? Why not add fries to it? Would we like to supersize it? Bundling like this traps us because we don't know where exactly to place value. When we face a bundle of this type we cannot easily value each of the individual components, because if we remove one item, it changes the whole price structure. If three items are each priced at $5, but together are only $12, which is the one that's overvalued at $5? Which one is the one we get on discount? Or are we getting a deal on all three? How much is a soda worth, at what size? And what about the value of the novelty cup?! Oh, I'll just take number one! Call my cardiologist.

If we identify bundles this way, we will quickly recognize that life is full of such bundles, many of which seem to be designed to confuse us. When we buy a home for $250,000, that's not the actual, total amount we'll spend, but it is the figure we rely upon. In practice, we pay a down payment, plus a monthly figure, for fifteen or thirty years, that includes some percentage of the principal plus interest at a rate that may or may not change. Then there's insurance and taxes, which will also change over time. And closing costs like appraisals, inspections, title search and insurance, agent fees, lawyer fees, survey fees, escrow fees, underwriting fees, and coming-up-with-new-fees fees. It would be difficult to separate each of those out to shop for the best bargain, so we lump them together and say we are purchasing a $250,000 house.

Of course, all service providers prefer to hide their fees within this large sum, to make these costs go unnoticed or, when we do notice them, to take advantage of our tendency to use relativity.

Or think about buying a cell phone. It's virtually impossible to compare a phone and its unique service plan to competitors' phones and plans. By design, each individual item is hard to value on its own: What are text messages worth compared to gigabytes of data? 4G networks, overage charges, minutes, roaming, coverage, games, storage, global access . . . What are they worth? What about the service and fees and reputation of the provider? How can we compare an iPhone on Verizon to an Android on T-Mobile? There are too many small, integrated elements to assess the relative value of each one, so we end up comparing the total cost of the phone and monthly service. If we can even figure those out.

Relative Success

The list of things that are affected by relativity extends beyond products like cell phones and ugly sweaters. Relativity affects our sense of self-worth, too. We have friends who attended some of the best schools in the country. By all reasonable measures, some of these friends are doing very well. Some, however, think of themselves only in comparison to their more "successful" top-tier colleagues, country club co-members, and golf buddies—and thus frequently feel like they aren't doing well. Jeff remembers quite vividly, and quite sadly, being at an exquisitely catered birthday party of a friend. While standing in the study of his five-bedroom, Park Avenue, doorman-building apartment, surrounded by supportive friends and a beautiful, healthy, and happy family, the birthday boy sighed and confessed, "I thought I'd be in a bigger apartment by now."

Objectively, he should have been celebrating his success. But, relative to a few other select colleagues, he considered himself a disappointment. Thankfully, as a comedian and writer, Jeff cannot compare himself to his banker friends. This allows him some perspective and allows him to be relatively happy with his life. Even more thankfully, Jeff's wife cannot compare him to a banker, though she does claim to know some funnier comedians.

The point is, relativity leaks into every aspect of our lives, and powerfully so. It's one thing to overspend on a stereo; it's quite another to lament our life choices. Happiness too often seems to be less a reflection of our actual happiness and more a reflection of the ways in which we compare ourselves to others. In most cases, that comparison is neither healthy nor good. In fact, our tendency to compare ourselves to others is so pronounced that we had to come up with a commandment not to covet thy neighbor's stuff.

In some ways, the concept of regret is itself just another version of comparison. With regret, we compare ourselves—our lives, our careers, our wealth, our status—not to other people, but to alternative versions of ourselves. We compare ourselves to the selves we might have been, had we made different choices. This, too, is often neither healthy nor useful.

But let's not get too deep and philosophical. Let's not worry about happiness and the meaning of life. At least, not just yet. Just take those emotions and store them away in a little box. Compartmentalize these things.

Like we do.

5

WE COMPARTMENTALIZE

Jane Martin doesn't hate her job. She just hates what she sometimes must do at her job. She's the events coordinator for a small state college, but now and then it feels like all she coordinates are rules, regulations, and how often she and her colleagues say no to each other. She needs approvals to get money from the activity fund or the general fund or the alumni fund. Every little item, from entertainment to tablecloths to transportation, must run through a hierarchy of budgetary paperwork. And it's not just college departments, the alumni groups, and the students who watch her mercilessly, ready to pounce on any slight mistake. It's also the state and federal rules. It's constant squabbling about finances and procedures because everyone needs a box checked next to their name. She loves putting on events. She hates worrying about paperwork.

At home, however, it's a different story. Jane is a detail master. She runs a tight ship, with a rigorous budget, and she loves it! She knows that each month her family can spend a certain amount of money on certain things. Two hundred dollars on entertainment. Six hundred on groceries. She sets aside money for home repair and taxes and medical care every month, even if she doesn't have those expenses. She actually puts cash for each category into labeled envelopes, so if she and her husband want to go to dinner, they have to see what's in the "dining-out" envelope to know if they can afford it. She doesn't let the family plan vacations too far in advance. At the end of each calendar year, if there's money left over in the home repair, taxes, or health expenses envelopes, she'll pool it together for a trip for the following summer. Using this approach, she's managed to save enough for some wonderful trips every year but one in the last ten—her daughter had to get knee surgery in 2011 after a soccer injury, so that burned up all the vacation funds.

Jane dislikes the month of October, because there are seven friends and family birthdays that month and she always burns through her gift envelope. This year, instead of getting her cousin Lou nothing, or dipping into the entertainment envelope to borrow money to get him a gift, she spent four hours making him a cake from scratch. He was excited to get the cake. She was exhausted.

WHAT'S GOING ON HERE?

Jane shows us an extreme example of *MENTAL ACCOUNTING*, another way we think about money that has little to do with actual value. Mental accounting can be a useful tool, but it most often leads to poor decision-making, especially when we're unaware we are even using it at all.

Remember fungibility? The idea that money is interchangeable with itself? A single dollar bill obviously has the same value as any other dollar bill. In theory, that's true. In practice, however, we don't usually assign the same value to every one of our dollars. The way we view each dollar depends on which category we first linked this dollar to—or, in other words, how we account for it. This tendency to place different dollars in different categories—or in Jane's case, envelopes—is certainly not the rational way to deal with money. But, given how difficult it is to figure out opportunity costs and real value, this strategy helps us budget. It helps us make quicker decisions about the ways in which we spend our money. That can be good, but by playing the mental accounting game, we also violate the principle of fungibility. We deny ourselves its benefits—we make things simpler and in the process we open ourselves up to a whole new set of money mistakes.

The idea of mental accounting was first introduced by Dick Thaler. The basic principle is that we operate in our financial behavior much like organizations and companies do. If we work for a large organization, like Jane's state college, we know that every year, every department gets its budget and they spend it as needed. If a department runs through its money early, too bad. The department chiefs won't get a new allotment until the start of the next year. And if they have extra money at the end of the year, everyone gets a new laptop or the holiday party might include fancy sushi instead of leftover bagels and donuts.

How does this approach to budgets apply to our personal financial lives? In our private lives, we also allocate our money to categories, or accounts. We generally set a budget for clothes and entertainment, rent and bills, investments and indulgences. We don't necessarily follow this budget, but we do set it. And much like companies, if we use all the money in one category, that's too bad; we can't replenish it (and if we do, we feel bad about it). On the other hand, if there's money left in a certain category, it's very easy to spend it. Maybe we don't go to the extreme of putting money in labeled envelopes like Jane, but we all use mental accounting, even if we're not aware of it.

Here's an example: Imagine we just spent $100 for a ticket to the hottest new Broadway show. It's a musical combining potty-mouthed Muppets, sassy superheroes, Founding Fathers, and high school hijinks. When we arrive at the theater on opening day, we look in our wallet and discover to our horror that we've lost the ticket. Luckily, we have another $100 bill in our wallet. Would we buy another ticket?

When people are asked this question, the vast majority say no. After all, they've spent the money on the ticket, the ticket is lost, and that's just too bad. Now, if we ask people to imagine that they went ahead and bought a replacement ticket, how much would they say that night of theater cost them? Most people say the experience cost them $200—the combined cost of the first and the second ticket.

Now imagine things went differently on the day of the show. We didn't buy a ticket in advance, but we're still just as excited about the production. When we arrive at the theater, we open our wallet and realize we lost one of the two crisp $100 bills we had in there. *Oh, no!* We are now $100 poorer. Luckily, we still have another $100 bill. *Oh, yes!* So, would we buy the ticket or just go home? In this case, the clear majority of people say they would buy the ticket. After all, what does losing a $100 bill have to do with not going to the theater? And, if like most people, we were to go ahead and get the ticket, how much would we feel we'd paid for it? In this case, the most common answer we get is $100.

Even though people react differently to those two situations, from a pure economic perspective, they are essentially the same. In both, there's a plan to go to a show and a lost piece of paper worth $100 (either a ticket or a bill). But from a human perspective there's a clear difference. In one case, the lost piece of paper was called a theater ticket; in the other case, it was currency—the $100 bill. How could the piece of paper make such a difference? How could this phenomenon cause us to go to the show in one case and to go home in the other? And how did we find such cheap Broadway tickets in the first place? (One hundred dollars? This theoretical world is quite affordable.)

Let's go back, for a second, to companies and their budgets. If we have a budget for theater tickets and we finish that budget (we use it on the ticket), we don't replenish it. Therefore, we do not get a new ticket. But if the money is lost from our wallet in general—rather than being spent on a specific item—we don't feel that it was taken from any particular budget category. Consequently, we don't see the need to punish any particular budget bucket. This means that there is still money in our theater-going account because the lost money came from the general expense account. So the loss doesn't stop us from enjoying the patriotic swearing puppet songs.

This mental accounting logic seems rather logical. So, what's wrong with this?

Accounts Deceivable

From a perfectly rational perspective, our spending decisions shouldn't be influenced by imaginary budget accounts, no matter how those accounts might vary in form, location, or timing. But they are.

We do this kind of mental accounting all the time. Think about some of the ways we keep our money in different accounts:

1. We put some money in low-interest checking accounts, while maintaining a balance on high-interest credit cards.
2. Jeff will sometimes bring his family along when he speaks or performs in interesting cities, like on a recent trip to Barcelona. When this happens, no matter how much he earns or how much the travel costs, he always overspends. It is easy to spend more of the money he gets for his performance, because he is getting and spending the money together. The growing earnings account overshadows the diminishing vacation expenses account, so all spending rules go out the window. In his mind, the money for each meal or attraction isn't coming from his family travel, education, or housing budget. It's coming from his speaking fee—every time. If they were just on a family trip, he'd be much more financially conscious or at least he'd ask more passive-aggressive questions, like, "Do we really need another glass of Cava?" (FWIW: The answer to this question is always "Yes. More, please.")
3. The entire city of Las Vegas is a great example of mental accounting. City tourism officials know we do this. They even have a marketing slogan designed to help us compartmentalize: "What happens in Vegas stays in Vegas." They encourage our

basest impulses, and we're more than happy to oblige. We go to Vegas and we put all our money into a mental Vegas account. If we win at the table games, great, it's a windfall. If we lose, no big deal, we already counted it as spent by putting it into that Vegas account. The truth is, we can put it in whatever mental account we want, and it's still our money; it just doesn't feel this way. Whatever happens to it while in Vegas—if we lose or win a few grand—that money actually does follow us home. It doesn't stay in Vegas. Neither do racy pictures posted on Instagram, so leave the phone in your room.

Gary Belsky and Thomas Gilovich retell the fable of the man who goes to play roulette with $5, starts an incredible run of luck, and at one point is up almost $300 million.[1] He then places one bad bet and loses all his winnings. When he gets back to his hotel room and his wife asks how he did, he says, "I lost $5." If this happened to us, we'd certainly feel like we'd lost more than $5, but we would probably not feel as if we lost $300 million. The $5 is all that ever feels like "our money"—what we started with that evening. We would categorize each dollar we gained that night, from the first one up to the 300 millionth, as "winnings." So, in this scenario, we may have lost $300 million from our winnings, but we would feel that we only lost $5 of our own money. Of course, we also lost the ability to communicate honestly with our spouse, but that's for a different book.

None of those scenarios makes sense when we consider that all the money being spent, saved, gambled, or drunk really comes from the same big pool of "our money." It shouldn't matter how we label the money, since in reality it's all ours. But—as we explained earlier—we do assign money to mental categories, and this categorization controls how we think about it from that point on. How comfortable we feel about spending it, on what, and how much we have left at the end of the month.

Mental Accounting: A Very Special Problem

Unlike most of the problems we discuss in this book, mental accounting is more complex than just "It's a mistake to use mental accounting." Mental accounting—like the others—is not a rational approach to money, but when we take into account the reality of our lives and our cognitive limitations, it *can* be a useful strategy. This is particularly true if mental accounting is used wisely. Of course, we don't often use it wisely, which is why the rest of this chapter exists. For now let's talk about why mental accounting is particularly unique.

Imagine there are three types of people: 1) the perfectly rational person—*Homo economicus*; 2) a somewhat rational person with cognitive limitations—he or she can determine the best decision if they have the time and mental capacity to figure it out; and 3) a somewhat rational person with cognitive limitations *who also has emotions*—that is, a human being.

For the perfectly rational person—all kneel before our robot masters!—mental accounting is unambiguously a mistake. In a perfectly rational world, we should treat money in one account the same as we treat money in any other account. After all, it's just money. Money is money is money. It's totally interchangeable. In the perfectly rational world we have an infinite capacity for financial computations, so it's a mistake to compartmentalize because it violates the principle of fungibility and denies us that major benefit of money.

For the person with cognitive limitations, with the real-life limits of our brain's capacity to hold and process information, mental accounting can, however, help. In the real world, it's extremely difficult to figure out the opportunity costs and multifaceted trade-offs of every single financial transaction. Mental accounting provides us a useful heuristic—or shortcut—for what decisions to make. Every time we buy something like a coffee, we can't reasonably think, "Oh, this could be a pair of underwear or an iTunes movie download or a gallon of gas or any of an infinite number of other purchases now or in the future."

Instead, we can use mental accounting to think of that coffee as part of our "Food" account. This way we just have to consider the opportunity costs within that account. This makes our thinking more limited but more manageable. "Oh, this could be half my lunch today or an extra coffee Friday afternoon." That simplifies the calculations. From this perspective, mental accounting is still not rational, but it is sensible, especially given our computational limitations.

When we compartmentalize for simplicity, we don't have to think about the whole world of opportunity costs every time we spend. That would be exhausting. We just need to think about our smaller budget—for coffee or dinner or entertainment—and the opportunity costs within it. It's not perfect, but it helps. In fact, once we recognize that mental accounting is not rational but can be useful, we can think about how to do more of it in a positive way.

That brings us to our third type of person, the ones with emotions and stress and annoyance and deadlines and *a lot of other things to do*! In other words: We, the *Real* People. While not as nearly impossible as figuring out the comprehensive opportunity costs of every transaction, constantly doing so even within smaller categories is, at a minimum, annoying. If we have to think about the pros and cons of our decisions every time we want to buy a specific item—coffee, gas, an app, this book—it's going to become a huge pain in the derriere (pardon our French). Much like how asking dieters to count every calorie often results in frustration, bingeing, and the counting of exactly no calories, the creation of complex budget categories often gets people to stop budgeting altogether. That's not the solution we want.

In fact, when people tell us that they have a hard time controlling their spending, we acknowledge that they could budget for everything, but we also tell them that it's likely to be so annoying that they'll just give up. Instead, we suggest they decide how much they want to spend on a broad category of "discretionary items": the things that they *can* live without, like special brew coffee, fancy shoes, or a night of drinking. Take that amount, on a weekly basis, and put it on a prepaid debit card. Now they have this category of discretionary spending with a new budget each Monday. The balance on the card will show how it's being used and the opportunity costs *within this general category*, and the opportunity cost of the decisions will be more apparent and more immediate. They can just look at the balance for discretionary spending. It still requires effort, but it's not as annoying as separate accounts for coffee, beer, Uber, and the digital version of this book. This is one way we can use mental accounting in our favor while recognizing the complexity and pressures of our real lives.

MORE SOLUTIONS TO COME

As you can see, mental accounting is a unique flaw in the way we think about money: In general we shouldn't engage in mental accounting, but since it simplifies life, we do. That, in turn, means that we should be aware of the mistakes we make when we do so. Acknowledging this shows how we can redesign the way we use money when we consider and embrace our money-spending nature.

We'll offer more tips like this—ways to take our flawed financial thinking into account and use it to our advantage—in the last section of this book. But now let's just continue exploring our money-based irrationalities. We'll place the rest of the solutions into a different literary section, or, you might say, a different mental account.

Out of Sorts

Our categorization of money affects how we treat it and how we use it, but we don't always have clear ways with which to categorize our money. Unlike a company, our lives aren't filled with office supplies and payrolls. We sort our money into different types of mental accounts, with different rules, depending upon how we get it, how we spend it, and how it makes us feel. Did we get this money from a job or from a lottery ticket found on the sidewalk? Or is it from an inheritance, embezzlement, or a career as an online gamer?

For instance, if we get a gift card for Amazon or iTunes, we will probably buy things we wouldn't normally purchase if that same amount had come from our paycheck. Why? Because a gift card goes into our gift account, whereas our hard-earned job money goes into a more protected, less frivolous account. Those accounts have different spending rules (even though, again, all of it is our own, fungible money).

A curious finding about the way we categorize money is that people who feel guilty about how they got money will often donate part of it to charity.[2] Let that sink in: How we spend money depends upon how we *feel* about the money. Yes—another hidden factor that influences how we compartmentalize our money is how it makes us feel. Do we feel bad when we get it because it arrived under negative circumstances? Do we feel it is free money because we got it as a gift? Or do we feel good, like we worked hard for the money—so hard for it, honey—so we deserve it?[*]

People are likely to spend something like their salary on "responsible" things like paying bills, because it feels like "serious money." On the other hand, money that feels fun—like $300 million in casino winnings—is likely to be spent on fun things, like more gambling.

Jonathan Levav and Pete McGraw found that when we get money that *feels* negative, we try to "launder" it. For instance, if we inherit money from a beloved relative, the money feels good and we are ready to spend it. But if we receive it from a source we don't like—in their experiment, it was the tobacco company Philip Morris—the money feels bad. So, to clean it of the negative feelings, we first spend some of it in

[*] Our next book will be about how, now that we've mentioned it, you'll never get the Donna Summer song "She Works Hard for the Money" out of your head.

positive ways, like buying textbooks or donating to charity, rather than selfish ones, like ice cream. Once part of the money was used for good, the money feels clean, and we feel perfectly fine spending the rest on more indulgent things like vacations, jewelry—and ice cream.

Jonathan and Pete call this *EMOTIONAL ACCOUNTING*. Emotional money laundering can take many forms. We might cleanse badly tainted money by first spending it on serious things like paying down debt, or on virtuous ones, like buying ice cream—for an orphanage. When we do something we think is good, it eliminates the bad feelings associated with the money, making us free to spend. This type of emotional money laundering is certainly not rational, but it makes us feel good.[3]

That's a fairly accurate statement about how we handle money in many situations: We don't handle it in a way that makes sense, we handle it in a way that feels good. (That probably applies to how we handle most things in life, too, but this is neither the time for philosophy nor the place for therapy.)

A Rose by Any Other Name Would Still Cost Us More

In some unfortunate ways, we act just like corporate accounting departments—like when we use accounting tricks to game the system for personal gain. Then we're like certain specific companies, like Enron. Remember Enron? The notorious energy company—the poster child for corporate cheating in the 2000s—made insiders obscenely rich by using fraudulent accounting schemes. Enron officials created offshore accounts to hide expenses and create phony income. They deceptively traded derivatives of basically fictitious products. Their entire accounting operation was "kept in check" by an auditing company that they themselves funded. They were cheaters. They were so good at it that they even started believing in the logic of their own fraudulent accounting approach.

Much of the buildup to the financial crisis of 2008 was generated by accounting schemes—by some in the financial industry making money from money itself, just moving it around, cutting it up, and selling it off. They skimmed from the top and shuffled funds between accounts when convenient, when profitable, and when it benefited them.

We perform similar accounting tricks on ourselves. We charge our credit card for different purchases and then quickly forget about them. We borrow from what we intended to save. We don't think about big bills when they're not in our monthly budget. We move money between savings and checking and rainy-day funds just so we can do something "special" with them. Most of the time, however, our accounting tricks don't cause worldwide economic meltdowns. Most of the time, they only melt our personal financial future. Most of the time.

Okay, maybe we're not as bad as Enron and its peers from the turn of this century, but we are shady with our mental accounting. We're easily led astray by emotions, selfishness, impulse, lack of planning, short-term thinking, self-deception, outside pressure, self-justification, confusion, and greed. We might consider those the Ten Financial Sins. Not Deadly Sins, but certainly not good.

And like the Enrons of the world, our mental accounting department is kept in check only by lazy auditors who don't want to think too much, love the pleasure of spending, and are burdened by an inherent conflict of interest. *We* are our own auditors. We are the fox guarding our own financial henhouse.

Imagine it's dinnertime and we're hungry. We ordered in last night and planned to cook tonight, but we didn't go shopping. Our budget says we shouldn't eat out, especially not at that hip new restaurant down the street. Sure, our friends are going out tonight, but we should whip something up at home and put the money we don't spend into a retirement account that'll earn compound interest until we're eighty. Then we'll be able to afford to eat out all the time. But we forget to ask ourselves, "What would Jane Martin or Moses do?" So we call the babysitter and an hour later, we're seated at the table, fancy cocktail in hand.

We promise ourselves we'll eat cheap and healthy. But look at this selection! We thought we'd have chicken, but that lobster in a wine-and-butter sauce just reaches out and wraps its succulent claws right around our eager throat. "Market price." Not bad; we heard it was a good year up in Maine. So we get the lobster and mop every last drop of the rich sauce with some thick slices of toasty bread. We also thought we'd survive on tap water, but we say, "Heck yes!" to a bottle of that fancy pinot. We really should skip dessert, but, ohhhhh—triple-glazed soufflé.

By the time the bill rolls around, we've gone way above and beyond the $6 or so that a bowl of pasta and an orange would have cost at home. We've violated our own dietary and financial accounting rules, but there's not a whistle-blower to be found.

We don't feel bad about eating and spending. After all, we have to eat something and we deserve a treat after a long week, don't we? Plus, after a little too much to drink, we lose the cognitive capacity to think about boring things like savings or paying our bills.

Even though it is irrational, mental accounting, just like corporate accounting, can be useful if used judiciously. Budget categories can help us plan our finances and control our spending. But, just like corporate accounting, mental accounting is not a panacea, because it still offers a lot of gray area. Just as some companies exploit loopholes with "creative accounting," so do we with our flexible spending logic. We mismanage our money when we don't use any categories, but even when we do use them, we then tweak the classification of our expenditures. We change the rules and we make up stories that fit our whims.

Mark Twain describes one such instance of creative manipulation of rules. Having limited himself to one cigar a day, he started shopping for bigger and bigger cigars, until he had each one made to such proportions that he "could have used it as a crutch."[4] Social scientists call this type of creative bookkeeping *MALLEABLE MENTAL ACCOUNTING*. We play with malleable mental accounting when we allow ourselves to classify expenses ambiguously and when we creatively assign expenses to different mental accounts. In a way, that helps us trick the account owner (ourselves). If our mental accounting weren't malleable, we'd be strictly bound by rules of income and expenses. But, since it is malleable, we manipulate our mental accounts to justify our spending, allowing us the luxury of overspending and feeling good about it.

In other words, even though we knew our budget shouldn't allow it, we found a way to make dinner work. Maybe we shifted the meal from the "food" to the "entertainment" account. Maybe we just decided that it is not our responsibility to send our kid to college. Essentially, we acted like a self-contained Enron, taking Wite-Out to financial plans to satisfy immediate desires. We won't go to jail for it, but we violated our own rules. We tore down the wall between food and entertainment and all hell—all that delicious, triple-glazed hell—broke loose.

Not only do we change how we use different categories, we also change the rules that define those categories themselves. When we have a not-so-great habit like buying lottery tickets or cigarettes, we often set up arbitrary rules for when we allow ourselves to purchase them. "I'll only buy the Powerball if the jackpot is more than $100 million." Of course, this rule is silly because the lottery is a bad decision no matter the jackpot size. It's like saying, "I'm only going to smoke cigarettes on partially cloudy days." But the rule makes us feel better about what we know is a poor choice.

Of course, we inevitably fudge these made-up rules whenever we can justify it—when our office pools money for lottery tickets or when we're standing in a long line at a checkout or when we are extra daydreamy or when the day has been difficult and we feel we deserve it. Since we're the ones who made the rules, and often the only people who know they exist, it's remarkably easy to change, amend, or override them with new rules without any repercussion. ("The $100 million minimum rule shall be suspended for all lottery purchases made while wearing brown slacks.") Our internal legislature is sure to approve, no matter the partisan rancor, no matter how little deliberation.

BAD MONEY CHASING GOOD MONEY

Let's say we *do* get a windfall, like a modest lottery win or a Barcelona speaking fee. Without thinking too much, we can easily spend it many times over, letting the good feeling of the indulgent, guilt-free bonus account bleed into our shrinking accounts. We splurge, telling ourselves that all of these purchases are covered by the windfall, even when we have long ago finished spending from that account. For instance, in Barcelona, Jeff justified several extra purchases (often sparkling wine, but not always!) by thinking of each of them as simple withdrawals against his speaking fee. It was easy, in the moment, to think of every single purchase as being the one special expense to celebrate his speaking gig. In reality, all of those single indulgences added up to a pretty large amount, but he never thought of it that way. At least, not until he was paying his credit card bill a month later. (More on credit cards to come.)

Malleable mental accounting also allows us to dip into our long-term savings for whatever present need or desire we might have. It allows us to spend on health care when an emergency presents itself. It allows us to make up entirely new budget categories on a whim; even worse, once we have this new line item, spending on it becomes easier in the future. Who knew there was a line item for "Celebrate Surviving Wednesday with Happy Hour" and that it repeated every week?

Sometimes when we do manage to save money in one way, we reward ourselves by spending on unrelated luxuries we wouldn't normally buy, even though the point of saving in one mental account isn't to spend from another. When this happens—which isn't all the time, but often enough—we're rewarding good behavior with bad behavior that directly undermines the good. Saving an extra $100 one week is a good start, but celebrating the saving by spending $50 on something we wouldn't have purchased otherwise—like a dinner or a gift—doesn't help our overall finances.

Another way we engage in creative accounting is known as *INTEGRATION.* This is when we rationalize that two different expenses are actually one by basically assigning the smaller expense to the same category as the larger one. This way, we can fool ourselves into believing we're suffering just one big purchase, which is less psychologically draining than one large *and* one small purchase.

For instance, we add our $200 CD changer to our $25,000 car purchase and consider it simply part of the car. Or we buy a $500,000 house and $600 worth of patio furniture so we can sit on our beautiful new back deck. We frame the whole thing as a house purchase, not separate house and furniture purchases. By combining purchases this way we feel we haven't incurred two losses—the house *and* the furniture—from two accounts—housing *and* home décor. It's just one. Or, after an exhausting day of shopping, we buy an expensive dinner . . . and then dessert . . . and then a drink at the local bar. And we lump all of these indulgences together into a mental account vaguely recognized as "Suckered In by the Holidays Again."

We also cheat on our accounting by misclassification. For instance, Jane didn't want to spend money on a gift for her cousin Lou, so instead she spent hours making him a cake. That time and effort has a value: It's four hours she could have been doing something else, from relaxing to visiting her family to even making money. Financially speaking, is her time worth more than the $15 she could have just spent on a picture frame for Lou? Probably (though there is, of course, emotional value in making a personal gift for family). Speaking strictly of money—which is Jane's focus—trading $15 for four hours of exhausting work is a bad decision, but one she made because of poor classification.

Our personal mental accounting rules are neither specific nor strictly enforced. They often exist as vaguely unrefined thoughts in our heads, so it is easy to find loopholes when we need or want to find them. As we've seen before, when given the choice, most of us will take the easy way out: We'll choose the most immediately tempting option, then use classification gymnastics to justify it without paying too much attention, even when the decisions we are making mean that we're cheating ourselves.

There is no limit to the effort people will make just to avoid thinking.

We're not bad people. Most of us are not consciously greedy, stupid, or ill-meaning by nature. We don't blatantly or recklessly violate our mental accounting rules, but we do use the malleability of the rules to justify monetary decisions that fall outside those rules.[5] Like cheating on a diet, we take advantage of our creativity and use it to justify almost anything pretty easily. After all, we deserve that ice cream cone since we had a salad for lunch earlier this week, right? And the ice cream truck is a local business to support, isn't it? And it's also only summer once a year, yeah? So let's treat ourselves! *Sprinkles!*

Timing Is Everything

You can't stretch time, can you? We try constantly. In fact, perhaps the most common way we cheat on our mental accounting comes from the way we think and misthink about time. Specifically, the time gap between payment for an item and our consumption of it.

One of the most interesting characteristics of the way we classify our financial decisions relates to the mental account into which we put a purchase, and the feelings we have about it, which often have to do with the amount of time between when we bought it and when we consumed it, rather than the actual value of the item. For example, Eldar Shafir and Dick Thaler studied wine—a wise and delicious choice—and found that advance purchases of wine are often thought of as "investments."[6] Months or years later, when a bottle of that wine is opened, poured, savored, consumed, and bragged about, that consumption feels free. No money was spent on fine wine that evening. Rather, the wine was the fruit of a wise investment made long ago. If, however, we were to have bought the wine that very day—or, heaven forbid, we were to drop and break the bottle—the purchase would feel like it came from *today's* budget. In this case, we wouldn't be patting ourselves on the back for a wise investment—because there was no time between purchase and consumption to establish it in a different category. In every wine-drinking situation—buy before and drink today; buy today and drink today; buy before and break today—we spend money on a bottle of wine, but depending on the timing of the purchase and the time gap between the purchase and the consumption, we think about the cost very differently.

What a bunch of self-deceiving little troublemakers we are. At least we're drinking wine while making trouble.

Timing isn't just important when it comes to spending money—it also matters in making it. What would salaried employees prefer: a raise of $1,000 per month or a bonus of $12,000 at the end of the year? The rational thing to do is to prefer $1,000 a month because if we get the money before the end of the year, we can save it, invest it, pay debt, or use it for our monthly needs.

However, if we ask people how they would use a $12,000 lump sum versus an additional $1,000 a month, most say they would spend the lump sum on something special to make themselves happier. That's because a lump sum payment would not arrive along with the usual monthly ebbs and flows of income and expenses—putting it outside of our regular account system. If, on the other hand, the money is received monthly, it would be categorized as salary—and most people would use it to pay normal expenses. Bonuses don't have this monthly time frame, so they can be spent on treats that we want but feel guilty about buying (which this chapter suggests might be wine and ice cream, but let's not judge).

More evidence of our preference for the fun of bonuses comes from the IRS—which is not an institution normally associated with words like "special" and "fun." Americans want tax refunds because getting money on April 15 feels like a bonus. We could set up our withholding so that by the end of the year we neither over- nor underpay our taxes and thus neither owe nor are owed anything in April. Instead, many of us choose to pay *too much* in taxes each paycheck—deliberately underpaying ourselves throughout the year—so that we receive an April bonus, aka the refund. A yearly bonus *from the government*, at that. Pretty special. Too bad we don't as easily part with our money for other, more productive causes.

Paying for Free

Those of us who live in a city and own a car know how expensive an urban vehicle can be. We pay higher insurance rates in the city. City driving is hard on cars, so maintenance costs are higher. We pay for parking meters, parking spaces, and totally unfair parking tickets. On top of that, city dwellers don't use our cars nearly as much as those who live in the suburbs. Rationally, many city dwellers should take taxis and rent cars for the occasional weekend adventures and trips to the suburban superstore. Those expenses would add up to be far less than the cost of owning a car. Nonetheless, whenever city folk use their cars—to shop, get away for the weekend, or to visit friends "in the 'burbs"—they feel like the trip costs them nothing. It feels as if they're saving money on the taxis and car rentals others must endure, and they're getting what is basically a free trip. This is because they paid for the trip with their regular, ongoing payments, but not directly at the time of the trip itself.

Similarly, with vacation timeshares, we pay a large up-front amount for the right to use a property anytime we want. For free! Well, yes, we pay nothing during the week we use the property, but we do pay—bigtime—usually once a year. But it feels free because the time of purchase and the time of use are different.

Accounts Payable

Mental accounting has an outsize impact on our money decisions. It directs and misdirects our attention and thinking about what to spend and not spend. But remember: It's not always bad. Given our cognitive limitations, sometimes mental accounting allows us to create useful shortcuts and maintain some sense of financial order. But in doing so, we often create loose accounting rules that can negatively influence our ability to assess value. This is particularly true when we separate—either by time, payment method, or attention—the pleasure of consuming something from the pain of paying for it.

Oh, you didn't realize that paying for things causes you pain? Well, hold on to your wallet and turn the page. . . .

6

WE AVOID PAIN

Jeff is married—sorry, guys—and, as it happens, his honeymoon experience was very instructive about how we think about our finances. Here is his romantic tale of love and money:

Anne and I found a place we'd wanted to go for a while—a nice resort on the Caribbean island of Antigua. We'd heard about this magical place from friends and it sounded like a great way to celebrate (and recover from) our wedding. The pictures looked beautiful, and, buried in the details of planning an event for a bunch of people we kind of, sort of knew, the thought of lying on a calm and boozy beach was irresistible.

We decided to buy an all-inclusive, advance-purchase package. We debated: The all-inclusive would be more expensive than the à la carte, pay-as-you-go option, and we would probably also eat and drink too much. But, after months of crash dieting to look good in our wedding duds, we went for it. It was appealing, in part, because it seemed so simple. Once we'd booked and paid for it, we could also cross an item off our seemingly endless to-do list. Who knew planning a wedding was so hard? I had thought it was just a matter of renting a tux and opening presents. Nope. You gotta do stuff like flowers, seating charts, and, of course, writing wedding vows. It's hard work.

We think wedding planning should be a mandatory first-date activity: If a couple makes it through that, then they can go see a movie. Otherwise, it won't work out. We are willing to bet that if starting with wedding plans was the standard courting process, there would be fewer incompatible couples. Marriage is hard!

Note: Not all of our ideas are good.

Anyway, our wedding was great. Lots of love, laughs, and a Ben & Jerry's ice cream wedding cake—highly recommended.

A couple of days later, we jetted down to Antigua and, after a billion hours of sleep, we really got into our vacation. Yes, we overate and overdrank and overeverythinged. There was so much to do. Like eating. And drinking. And eating and drinking. A hearty breakfast, some Bloody Marys, a seafood lunch, coconut-based cocktails, naps, some kind of rum drink, dinner, fine wine. And dessert. We had lots of dessert. I mean, they just rolled the dessert tray out there every night. What could we do? At home, we wouldn't indulge, but, you know, we were pretty sure all the extra calories wouldn't be allowed back through customs.

We managed to fit in some activities, too—swimming, tennis, sailing, and snorkeling. We even went on a few excursions that we ended up cutting short (whether that was due to our desire to read in depth about the history of Antigua or to not enough rum, I will leave to your imagination). While we felt a little spoiled, we also felt like we deserved to treat ourselves. The only time we felt guilty about indulging was when we periodically left about half a bottle of good wine undrunk. Not that we only *had* half a bottle ourselves; the half left behind was usually our second or third bottle of the evening.

It turned out that one of the unexpected joys from our prepaid all-inclusive vacation was that the resort posted the prices for everything everywhere. Labels adorned food, drink, and beach towels. The prices were plastered on beach chairs. They confronted us on boat rides and island trips. At first we thought it was tacky, but then we began enjoying being reminded of all the free food and fun we were having and all the money we were saving.

It was an escape from reality. From wedding planning, wedding having, wedding family. We were fat and drunk and sunburned.

Then, in the middle of our stay, it started to rain. It rained and rained. For three straight days.

Normally, this would be a bummer. You want to lie on the beach on your honeymoon, right? But sometimes, when life gives you lemons, you make lemon-rum punch.

We relocated to the resort bar. We tried every drink they had. Some we liked; some we left unfinished. All this merriment helped us befriend other honeymooning couples who were also taking refuge in the bar. They were good people, some of whom we still talk to regularly and visit from time to time, though time and rum have blurred our memories of those rainy days.

One couple from London—let's call them the Smiths—arrived right when the rain started. They declined to join us on our "try every drink" challenge. Instead, they sipped down every drop of each concoction they ordered, even when their faces showed no particular pleasure from the drink. (Diagnosis: not enough rum.)

After the days of rain ended, we'd catch up with the Smiths on the beach or at a restaurant—but only for dinner. They often skipped breakfast and just had a big evening meal. They didn't drink much, even though they joked a lot about pub nights back in jolly ol' England. A couple glasses of wine at dinner, hardly anything on the beach. And they seemed to argue a lot. Now, we're not ones to judge—but we judged. Turns out they'd chosen the à la carte plan and were having some differences of opinion about what to spend their money on. It was understandable, kind of: The drink prices and activities fees weren't cheap, and just talking about what to do and what to spend added tension to their new marital bliss.

We checked out of the resort on the same day as the Smiths. As we hopped on the airport shuttle, we saw them sorting through a nineteen-page bill with the resort staff. It was a sad way to end our time together, especially since they missed that shuttle and almost missed their flight.

Missing a flight might have been a bit of a blessing, though. Getting stranded in Antigua? Our luck was to get stranded in Miami. It's a lovely town, but very few places are great on a short, unexpected visit. We were transferring between flights, and first an equipment issue and then an approaching tropical storm kept us grounded for a couple of nights. The airline offered to put us up in a hotel, and we accepted. We could have upgraded to a nicer location but decided it wasn't worth the extra $200. The place we stayed in was dingy and dirty and not in a great neighborhood, but we figured we'd just try to enjoy this little surprise. Neither of us had spent any time in Miami, so why not give it a shot for thirty-six hours?

We went right to bed, no partying, and in the morning popped into a local's favorite place for breakfast and shared a big omelet. I wasn't hungry enough to eat my own, and $15 seemed like a lot to spend on a few bites. It was pretty good. We went to the beach but didn't rent boats or water skis or umbrellas. We just sat and relaxed, which was nice. We could see the big storm on the horizon. Lunch was another share, and then we made plans for dinner and a show.

We went to a good restaurant, a place with a great view of the not-yet-stormy ocean. We filled up on bread, skipped the appetizers and salad, and had an entrée each. No wine. We did have a couple of cocktails each, but no dessert. We'd had enough sugar for a lifetime. (The prediction that customs would reject our extra calories proved false, sadly.) I was still a little hungry after, but figured I'd get a snack at the show.

Except that we didn't go to a show. There was a local calypso band playing at some hip new club, but by the time we got there, the only tickets available cost $35 each. That seemed pricey for a band we'd never heard of, so we took a nice walk back to the hotel. Then it started to rain. A lot. Tropical storm rain. We ran back to our room, slammed the door, and hopped into bed. Pulled out some books and read till we passed out. A nice, simple day.

When we finally got home, the evil long-term parking place overcharged us by a day, so I had to argue with them about that. We got home late and had to go right to bed so that we could wake up on time the next morning and go straight to work. A bad ending to a good trip. But isn't this the story of life?

Later that week, our friends wanted to hear all about our trip, and we were excited to tell them. So we all got together for dinner at a nice restaurant. It was fun—good to see everyone and great to be told how tan we were (it's the simple things in life). Then the bill came, and despite my best efforts, I could not help but point out that we—in an attempt to detox, perhaps—hadn't had any of the champagne or fancy wine that our friend ordered. There was some discussion about who should pay what, and in the end, everyone just looked at the bill and paid for their own items.

I asked the server if she'd accept payment in seashells and suntan. She didn't laugh. I gave her my credit card.

It was an unpleasant ending to a good evening out. But isn't this the story of life?

HAPPY ENDINGS

The end of an experience is very important. Think of closing prayers at religious services, dessert at the end of a meal, or goodbye songs at the end of summer camp. Ending on a high note is important because the end of an experience informs and shapes how we reflect back on, remember, and value the entire experience.

Donald Redelmeier, Joel Katz, and Daniel Kahneman studied how the conclusion of a colonoscopy (the ultimate "end-of-our-end") influences patients' memories of the whole procedure.[1] For some patients they used the standard way to end the procedure, while for others they added a five-minute component at the end. The addition was time-consuming but less painful. When doctors used the longer procedure with the less painful end, patients viewed the overall colonoscopy experience as less unpleasant, even though overall the ordeal had the standard procedure and then some more.

Of course, vacations are exactly nothing like colonoscopies—but the idea that the ending is important applies here as well. We often end vacations on a low note, with things we hate most: paying the hotel bill, shuttles, airports, taxis, suitcases, laundry, alarm clocks, and returning to work. Those ending activities can color how we view the vacation as a whole and paint it in a less positive way.

Our memory of a vacation—even one with three days of rain—would be better if we had a happier ending. How might we do this? We could "virtually" end the trip before we get into the unpleasant stuff by, for instance, celebrating the end of the trip the night before we check out. When we do that, we psychically place the packing, airport, and travel experience into the "regular life" bucket rather than the "end of vacation" one. We seal the trip in a box and keep the hassle outside it.

Another solution would be to prolong the trip. After we get

home and deal with reentry into everyday routine, we can make time to talk over memories and experiences, look at the pictures, and write some notes, all while the journey is fresh in our minds. Spending time savoring the vacation brings the experience into our regular lives and this, too, can give us a softer ending.

Finally, we could improve our vacation if, at the end, we remember that it was better than a colonoscopy.

WHAT IS GOING ON HERE?

Jeff's honeymoon experience shows us the many manifestations of the *PAIN OF PAYING*. The pain of paying is, as it sounds, the idea that we experience some version of mental pain when we pay for things. This phenomenon was first proposed by Drazen Prelec and George Loewenstein in their paper "The Red and the Black: Mental Accounting of Savings and Debt."[2]

We're all familiar with physical and emotional pain: a bee sting, a needle prick, chronic aches, and a broken heart. The pain of paying is what we feel when we think about giving up our money. The pain doesn't come from the spending itself, but from our thoughts about spending. The more we think about it, the more painful it is. And if we happen to consume something while thinking about the payment, the pain of paying deeply colors the entire experience, making it far less enjoyable.

The term "the pain of paying" was based on the feeling of displeasure and distress caused by spending, but more recently, studies using neuroimaging and MRIs have showed that paying indeed stimulates the same brain regions that are involved in processing physical pain. High prices stimulate those brain mechanisms with higher intensity, but it's not just high prices that cause pain. Any price does. There is a pain we all feel when we give up something.[3]

No Pain, No Pain

When we experience any pain, our first instinct is to try to get rid of it. We want to ease our pain, to control it. When we see pain coming, we flinch, we duck, we avoid it. We do that with the pain of paying, too. The trouble is, the way we often try to escape the pain of paying causes even more trouble in the long run. Why? Because we run from painful spending to painless spending, without regard for other, more important factors. This pain avoidance does not help our money trouble. It helps us avoid the pain right now, but often with a higher cost in the future.

Avoiding pain is a powerful motivator and a sly enemy: It causes us to take our eyes off value. We make faulty decisions because we're focused on the pain we experience in the process of buying, rather than the value of the purchase itself.

Pain hurts, but it is also important. Pain tells us something is wrong. A painful broken leg tells us to get help. The pain of a burn tells us not to touch fire. A rejection by Megan F. in seventh grade teaches us to be cautious with girls named Megan. Sorry, Megan H.

Now, a baby who touches a stove feels pain, over time he understands what's causing it, and eventually he learns to stop touching stovetops. So, too, we should learn what's causing us pain and avoid it. Do we do that? Do we stop doing painful things or do we just numb the pain so we can keep doing the painful things, pain-free? What do you think, Seinfeld?

> There are many things that we can point to that prove that the human being is not smart. The helmet is my personal favorite. The fact that we had to invent the helmet. Now why did we invent the helmet? Well, because we were participating in many activities that were cracking our heads. We looked at the situation. We chose not to avoid these activities, but to just make little plastic hats so that we can continue our head-cracking lifestyles. The only thing dumber than the helmet is the helmet

law, the point of which is to protect a brain that is functioning so poorly, it's not even trying to stop the cracking of the head that it's in.

—Jerry Seinfeld, *I'm Telling You for the Last Time*

The pain of paying should get us to stop making painful spending decisions. But instead of ending the pain, we—with the "help" of financial "services" like credit cards—devise ways to lessen the pain. Using credit cards, e-wallets, and automatic bill-pay is the equivalent of putting on little "financial helmets." Like bad doctors, we treat the symptom (the pain) but not the underlying disease (the paying).

This is one of the big mistakes that influence the ways in which we evaluate our money decisions.

The pain of paying is the result of two distinct factors. The first is the gap between the time when our money leaves our wallet and the time we consume the good for which we've paid. The second factor is the attention we give to the payment itself. The formula is: Pain of Paying = Time + Attention.

So, how do we go about our lives avoiding the pain of paying and how does that avoidance affect the way we value money? Well, we do the opposite of that which creates the pain. We increase the time between payment and consumption and we decrease the attention needed to make payment. Time and attention.

As for Jeff's experience, he and his lovely, patient, kind, out-of-his-league wife (are you reading this, honey?) paid for their honeymoon well in advance of the trip. When they wrote that big check, they undoubtedly winced. But by the time they arrived in Antigua, the payment and its associated pain were far in the rearview mirror. Every experience, every delight, every drink felt free. As they ordered another bottle of wine or took out a sailboat, they didn't have to think about money or whether the thing was worth it or not. They had already made their financial decision. They could just act on their whims, desires, and impulses—which they did. In fact, seeing the high à la carte prices that they *didn't* have to pay made them feel even better: In the moment, it felt like they were getting things for free.

The Smiths, on the other hand, experienced the pain of paying regularly during their stay. Every time they wanted to do something—drink, eat, swim, snorkel—they had to pay for it, feel the associated pain of paying, and experience the reduction in the fun that resulted from that pain. They didn't have to count out bills, per se, but they had to weigh the costs and benefits, charge the bill to their room, contemplate a tip, and so on. Even small items incurred an associated payment, and therefore an associated pain. Admittedly, the relatively small amount of attention they had to give to signing for tropical drinks at a Caribbean resort is probably the textbook definition of "first-world problems," but it was noticeable nonetheless. The Smiths were constantly dealing with the pain of paying, and it showed in their tension and bickering. "Till death do us part" seemed to be approaching quickly.

When Jeff and his new bride got stuck in Miami, they were still on their honeymoon—still in a relatively exotic location, in some ways. It was an unfamiliar place, they were traveling, they had airports, hotels, beaches, and all the fixings of a planned vacation. So they were willing to be a little cavalier with their spending, trying out things they weren't sure about. Their hotel was paid for, so they felt like they had some bonus money they could afford to spend (mental accounting). But it wasn't the same as having prepaid for everything. They still had to take out their wallets and fork over some cash or use their credit cards. They had to make some effort to pay and they had to give some attention to the money leaving their bank account. So, in Miami, they showed some restraint and didn't follow their every whim. They didn't go to the show they weren't sure about or order too much booze. They were more frugal than in Antigua. Bad news for the economy of Florida's coastline, good news for the size of Jeff's waistline.

When they got home, they became even stingier: They were feeling the pain of paying, in all its power. They were back to normal life, no longer under the mental account of their honeymoon. At the restaurant with friends, they were confronted with the burden of paying for someone else's wine right after having spent many thousands on the wedding and honeymoon. The pain of paying made them cranky. So, to ease their pain a little, they used their credit card. As we shall see, whipping out that piece of plastic didn't hurt as much as parting with cash would have.

SOME LIKE IT HOT

When we eliminate the pain of paying, we spend more freely and enjoy consuming things more. When we increase the pain of paying, our spending goes down as our control goes up. Should we always increase or decrease the pain of paying? Of course not. There's a time and place for everything.

There are certain experiences, like a honeymoon, that happen only once—or twice, or (if you're a politician) three times maximum—and these are very special occasions. In this case, we would argue that it's a good thing to reduce the pain of paying and just enjoy our onceish-in-a-lifetime experience. But in our daily lives, when we do things over and over and over, maybe there are categories for which we should increase the pain of paying. Buying lunch, grabbing trashy magazines at the supermarket checkout, getting a pricey smoothie after working out—these are things we can reconsider without ruining a priceless moment.

The point is, we can increase or decrease the pain of paying that we feel at any time, for any transaction. But we should do so deliberately, based upon how much we want to enjoy or limit our spending, rather than just letting it increase or decrease without our knowledge or control.

Time Keeps On Ticking, Ticking, Ticking . . . into My Wallet

When consumption and payment coincide, enjoyment is largely diminished. When they are separated, we don't pay as much attention to the payment. We sort of forget about it, and as a consequence, we can enjoy our purchases much more. It's as if we have a guilt tax that hits us every time we pay for something, but its effect on us is temporary, and confined to the time when we're paying—or thinking about paying.

There are basically three types of times we can pay for a product or service: before we enjoy it, as Jeff did with his honeymoon; during consumption, like the Smiths were doing; or after, like paying for that return-home dinner with a credit card.

Consider the timing aspect of an experiment run by Jose Silva and Dan:

Undergraduate students were paid $10 to sit in a lab in front of a computer for forty-five minutes. They could sit there and do nothing and leave with all $10, but they also had the option to buy entertainment for a low price. There were three categories of information that the students could view online: cartoons, the highly desirable category; news and science articles, the second-most desirable; and the third, the undesirable category, which was—you guessed it—cultural studies articles on postmodern literature. They could examine any piece of information they wanted, for a price. All the while, the computer kept track of their viewing and charged them three cents for each cartoon and half a cent for each news or science article. They could read as much of the postmodern lit as they wanted to for free.[4]

A NOT-SO-SIMPLE MISUNDERSTANDING

Are you a fan of postmodern literature? Do you even understand postmodern literature, or at least want people to think you do?

Then you should visit a wonderful website called the Postmodernism Generator (www.elsewhere.org/journal/pomo/). It randomly creates "postmodern" essays by pulling some quotes and throwing around names like "Foucault," "Fellini," and "Derrida." The site makes it so we feel like we understand each sentence as we read, but then, as we continue along, we realize we haven't understood anything. That's the feeling many people get from postmodern literature.

We considered using the Postmodernism Generator to write this book. Who knows? Maybe we did.

In addition, the method of payment was set differently for different groups. In the postpayment group, participants were told that the amount would be deducted from their payment at the end of the session, like a bill at the end of the month. In the prepayment group, the situation was like a gift certificate: The participants got the same $10, but all the money was placed in an e-wallet account that they could use for reading online material. This group was told that at the end of the experiment they would get all the cash that was left in their account. Finally, the third group was in the micropayment condition: These participants were charged every time they opened a particular article. Every time these participants clicked on a link, we asked them, "Are you sure you want to pay half a cent for this article?" or "Are you sure you want to spend $0.03 on this cartoon?" If they clicked "OK," they were charged immediately. Their remaining balance was always shown at the top of their screen. (Jeff often wonders where Dan finds so many students willing to participate in these experiments, and if he can have their contact info to "experiment" with them painting his house and babysitting his kids.)

Importantly, participants across the conditions paid the same amount of money for the pieces they were reading. Furthermore, across all the groups, they didn't spend very much (the price per item was low). However, there were big differences in spending based upon when participants were thinking about the payment.

When the money was placed into participants' entertainment account at the beginning of the study—in other words, the prepayment condition—the average participant spent about 18 cents. When they paid at the end of the study, like a regular bill (paying after), average spending dropped to 12 cents. This tells us that having the money in an account dedicated to a particular activity influenced our participants to spend more. Fifty percent more, in this case. The most impressive effect was on how much they spent in the micropayment condition, where they were forced to think about the payment every time prior to purchasing (paying during). In this condition, the average participant spent just 4 cents. On average, participants in this condition viewed one cartoon and two science articles and spent the rest of the time reading cultural studies—painful, but free. The combination of these results suggests that moving from paying after to paying before changes our choices. And, most important, when the payment is extra salient, we dramatically change our spending patterns. In short, because of the pain of paying, we're willing to pay more before, less after, and even less during consumption of the very same product. The timing of payment truly matters. It can even get us to read postmodern literature.

We don't want to pile on postmodern literature—it undoubtedly has some value, to some people, somewhere—but we should note that the participants in the study did not enjoy reading it and, in fact, they told us that they preferred the sounds of nails on a chalkboard to our version of postmodern literature. That means that the free activity—postmodern lit—caused the least amount of pain *of paying*, but the highest amount of pain *of consumption*. People enjoyed the experience of consuming postmodern lit much less than they enjoyed the experience of the cartoons. But by trying to avoid the pain of paying for the cartoon, the participants created the pain of consuming the postmodern lit. Those in the pay-as-you-go condition could have spent 12 cents instead of 4 cents and they could have had a much better overall experience for the forty-five minutes of the experiment, but the pain of paying is so powerful that it prevented them from doing so.

Similarly, imagine we're on a pay-as-you-go honeymoon. Our concierge offers us a nice bottle of champagne to drink on the beach at sunset, but because we're so annoyed by all the charges piling up and the asking price of the bottle, we decide to stick with tap water. Yes, we avoid the pain of paying for overpriced champagne, but we also avoid the pleasure of drinking champagne during a onceish-in-a-lifetime honeymoon sunset.

When paying as we go, we may now find it challenging to balance the pain of paying against the pleasure of consumption. As the Postmodernism Generator tells us Foucault said, "Life isn't easy, my man."

Paying Before

When Jeff paid for his honeymoon in advance, he consumed more and enjoyed it more than if he had paid for everything during or after the trip. He may have even paid more overall, and still his joy was higher. This pattern has not escaped the attention of some businesses. Prepayment has become trendy. Fancy restaurants like Trois Mec in Los Angeles, Chicago's Alinea, and New York's Atera are now encouraging customers to prepay for meals online.

But prepaying isn't just a trend, it's all around us. We buy Broadway tickets, airfare, and Burning Man passes well before we use them. Heck, you paid for this book before you consumed it, rather than waiting to finish the last page (at which time you'll likely want to send us a thank-you note with a substantial tip).

If we pay for something before consuming it, the actual consumption of it feels almost painless. There is no pain of paying at that time, nor any worrying about paying in the future. It is a pain-free transaction (unless it's the purchase of something that causes physical pain, like rock climbing, boxing lessons, or a dominatrix—but this is a family book, so let's move on).

Amazon.com relies on shifting the cost of shipping to prepayment with their yearly Prime membership, which costs $99 but promises free shipping throughout the year. Of course, it's not really free shipping—we've paid $99—but as we consider each purchase throughout the year, there is no additional pain of paying associated with each shipment. It feels free at that time, especially since Amazon slaps a brightly colored "FREE 2-DAY SHIPPING WITH PRIME" sticker right by the price. It feels as if we almost *have* to buy more, because we're getting such a great deal! And the more times we buy from Amazon, the cheaper, the "more free," each online shopping spree becomes. What a deal!

Imagine that we're going on a weeklong African safari that will cost us $2,000. We have two ways to pay for this adventure. We can pay for the whole trip four months in advance or pay in cash the moment we finish the safari. If asked which form of payment is more economically efficient, we would clearly answer that it's paying at the end, once services have been rendered. If nothing else, the money could be accumulating interest for those four months. But what about our enjoyment of the trip? Under which of those payment options would we enjoy the safari more, and in particular, under which would we enjoy the last day of the safari more? If we're like most people, we would enjoy the safari much more if we paid for it in advance. Why? Because if we paid for it on the last day, the last few days of the safari would be filled with thoughts like "Is this worth it?" and "How much am I enjoying this?" By having these thoughts constantly rattling around in our heads, our enjoyment of the entire experience would be vastly diminished.

Prepaying is also an inherent part of experiences such as gift cards and casino chips. Once money is put into a gift card for Starbucks or Amazon or Babies "R" Us, we put that money into spending categories—that is, once a $20 bill has been traded for a Starbucks card, that $20 has been allocated to lattes and scones, not, say, Coke and Chinese food. Moreover, once the money has been allocated to that category, we feel as if payment has already been made. We're not using our own money for anything, and as a consequence, we feel guilt-free while we spend it. We might normally just get a small coffee with our own money, but when spending from a gift card, we splurge on a Venti Soy Chai Latte and a biscotti. After all, it's free, right? We feel no pain spending a gift card because the feeling it evokes is nothing like the feeling we have when we spend cash.

It might seem obvious to say this, but we all like consuming things *and* we all dislike paying for them. But, as Drazen and George found, the timing of the payment matters a great deal, and we feel better consuming anything that we have already paid for.[5]

Pay During

How does paying for something while we're using it affect the pain of paying and our sense of value?

Imagine buying ourselves a fun little sports car as a retirement/midlife-crisis gift. We do it with a loan, incurring monthly payments. As it was intended to do, the car drives great and helps us forget our impending mortality and some of our poor life choices. However, we find we have less and less time to drive, and slowly even the thrill of the drive begins to wear off. Our monthly payments remind us of what was actually a rash and expensive purchase, one that's becoming harder and harder to justify. So we pay off the whole loan. Making that large, one-time payment is certainly painful, but it provides some relief from the monthly pain of regular payments and the associated guilt. It even restores some of the pleasure of zooming around with the top down. We stop being worried about the payments each month and begin to enjoy the car, even when we don't get behind the wheel that much.

Paying for things while we consume them not only makes us more acutely aware of the pain of paying, but it also diminishes the pleasure of consuming. What if a restaurant owner found out that, on average, people take 25 bites and pay $25 for a meal? This comes out to a dollar per bite. One day, the owner decides to have a 50 percent off promotion and charge 50 cents per bite. He then goes a step further and says, "I will charge only for the bites you take! The bites you don't take? You don't have to pay for those." As our food is served, the waiter stands next to us and makes a little mark on a notepad every time we take a bite. And when we're finished, the waiter rings up our check, charging us 50 cents per bite, and only for the bites we took. This is certainly a recipe for a very economical meal. But how much fun will it be? Doesn't seem like much fun at all, does it? Dan once brought pizza into his class and charged the students 25 cents per bite. What was the effect? Huge bites. His students, trying to avoid the pain of paying, thought they found a workaround by taking extra-large bites. Of course, they suffered while they ate, with clogged throats and messy faces, so it wasn't much of a bargain and it certainly wasn't a pleasure. More generally, pay-per-bite is often not a great way to pay, because it makes the dining experience incredibly unpleasant. That said, it might be an ideal way to approach dieting because the unpleasantness of eating will overwhelm the enjoyment. Not to mention that counting bites might be easier than counting calories.

One business-world example of how painful it can be to have payment coincide with consumption can be found in an examination of what happened when payment and consumption were actually separated by a little company called AOL. Millennials, if you're not sure what AOL is, google it.

In 1996, AOL president Bob Pittman announced that the company planned to replace AOL's two payment plans—$19.95 for twenty hours of usage plus $2.95/hour after that, or $9.95 for ten hours and $2.95/hour after—with a $19.95 flat rate for unlimited access. AOL staffers then prepared for the changes in the number of hours that their users would connect to their servers as a result of this price change. They looked at the distribution of how many people were using the service close to the threshold of ten hours and twenty hours, and estimated that the new plan would spur some customers to start using the Internet more frequently. They also assumed that most people would continue to use the service as they were, unless they were close to their hourly thresholds. When they were making these calculations, they believed that if a customer was using the Internet for only seven hours under the old plan, he or she couldn't possibly want to use it for much more after. Taking these assumptions into account, they increased their available servers by a few percent. Surely now they were prepared for the dawn of unlimited-access pricing, right?

Wrong. What actually happened was that the total number of hours people were connected more than doubled overnight. AOL was, of course, completely unprepared for this. It had to seek service from other online providers, which were happy to comply (and quite happy to charge AOL an arm and a leg for their services). In defense of his blunder, Pittman said, "We are the largest in the world. There is no historical precedent to consider. Who would have thought that they would double. . . . It's like a television station doubling its rating."

But could AOL data geeks really not predict this? If the AOL team had examined aspects surrounding payment and the pain of paying, they would have realized that when consumption and payment coincided, and when customers see at the top of their screen a clock counting down their remaining time—as with the old plans—it is hard not to think constantly about how much time is left and how much it would cost if they went over. In doing so, their enjoyment decreased. So the moment that the counter showing the remaining time to the end of the plan (10 or 20 hours) was eliminated, the pain of paying disappeared as well. So people were far more likely to use and enjoy the service for longer periods of time. Much longer.

The pain of ongoing, simultaneous payment isn't necessarily bad. It just makes us more acutely aware of our spending. Energy is an interesting example. When we fill up our car with gas, we watch the dollars spin by on the gas pump. Aware of our spending, we feel the pain of paying and perhaps contemplate buying a more efficient vehicle or finding a carpool group. But at home, the energy meter is usually outside or hidden. We rarely look at it. Moreover, the bill for the usage in any one day or week doesn't come for a month or more. And then it is often deducted directly from our checking account. Thus it's impossible to tell what we're spending at any one moment. So we are not as aware of our spending and we do not feel the associated pain. Perhaps there is a solution to our home energy use and overuse? (Spoiler alert: We will discuss this more in part 3.)

PAYING AFTER

Ah, the future. To understand how future payments—paying for something after we consume it—affect the pain of paying, we need to understand that we value money in the future less than we value it right now. If we were to have the option of $100 right now or $100 in a day, or week, or month, or year, most of us would choose the $100 right now. Money in the future has a discounted value. (There are countless studies about all the irrational ways we discount future outcomes.[6]) When we plan to pay in the future, it hurts less than when we pay the same amount now. And the further into the future we pay for something, the less it hurts now. In some cases, it feels almost free right now. We're not paying until the great, unknowable, optimistic future, when we may be a lottery winner or a movie star or inventor of the solar-powered jetpack.

CREDIT WHERE CREDIT IS DUE

This is one of the evil geniuses of credit cards: The main psychological force of credit cards is that they separate the time that we consume from the time we pay. And because credit cards allow us to pay for things in the future (when exactly is our payment due?), they make our financial horizons less clear and our opportunity costs more blurry, and they lessen our current pain of paying.

Think about it: When we pay for a restaurant meal with a credit card, do we really feel like we're paying right now? Not really. We're just signing our name; the payment will be sometime in the future.

Similarly, when the bill comes later, do we really feel like we're paying? Not really. At that point, we feel like we already paid at the restaurant. Not only do credit card companies employ the illusion of time shifting to alleviate the pain of paying, but they do it *twice*—once by making it feel like we are going to pay later and once by making us feel like we already paid. This way they enable us to enjoy ourselves, and spend our money, more freely.

Credit cards capitalize on our desire to avoid the pain of paying. And that has given them the power to shift the way we perceive value. With easier, less salient payment and the shifting of time between payment and consumption, credit cards minimize the pain of paying we feel at the moment we buy something. They create a detachment that makes us more willing to spend. As Elizabeth Dunn and Mike Norton noted, this detachment is not just about how we feel in the moment; it also changes how we remember the purchasing experience in a way that "makes it harder to remember how much we've spent."*[7] For example, if we go to the store and buy socks, pajamas, and an ugly sweater, the moment we get home, we're less likely to remember the amount of money we spent if we used a credit card than if we used cash. Credit cards are like memory erasers from a science fiction movie, but they live in our wallets.

Studies have found not only that people are more willing to pay when they use credit cards,[8] but also that they make larger purchases, leave larger tips, are more likely to underestimate or forget how much they spent, and make spending decisions more quickly. Furthermore, just displaying credit card paraphernalia like stickers or swipe machines—simply bringing credit cards and their "benefits" into our consciousness—also generates all these credit-card-influenced behaviors. That is not a typo: One study, way back in 1986,[9] found that just

* They also noted studies showing students underestimating their credit card bills by 30 percent and MBA students bidding twice as much on products when using credit cards.

putting credit card schwag on a desk induced people to spend more money.

In other words, credit cards—and even just the *suggestion* of credit cards—influence us to spend more, more quickly, more carelessly, and more forgetfully than we would otherwise. In some ways, they are like a drug that blurs our ability to process information and act rationally. While we don't drink, snort, or smoke credit cards—at least, not yet—their effect is deep and worrisome.

Credit cards also make us value purchases differently. They seduce us into thinking about the positive aspects of a purchase, in contrast to cash, which makes us also consider the downsides of the purchase and the downside of parting with our cash. With credit card in hand, we think about how good something will taste or how nice it will look on the mantel. When we use cash, we focus more on how fat that same dessert will make us and how we don't have a mantel.[10]

Same product, same price, but valued totally differently just based upon how we pay, how easily we pay, and how much pain it causes.

She Works Hard to Spend Money

The power of credit cards lies not just in temporal shifting—altering the time between pleasure and payment—but also in reducing the attention it takes to pay. The less attention, the less pain, the more we value something without cause.

A simple swipe of a card is easier than getting out our wallet, observing how much money we have, grabbing some bills, counting, and waiting for change. When we use cash, we actually think about, notice, touch, grab, remove, sort, and count the money we're spending. In the process, we feel the loss. With a credit card, that loss is not as vivid and not as visceral.

Credit cards also make payment easier and less painful by consolidating a month's purchases into one simple bill. Credit card companies are aggregators, putting all our purchases together—food, clothes, entertainment, etc.—into one lump sum. We accumulate a balance and, as a result, charging a little more for another purchase doesn't seem to hurt because it doesn't change the overall amount we owe the credit card company by much.

As we learned earlier in our chapter on relativity, when an amount—say, $200 for dinner—is put in the context of a larger amount—say, a $5,000 monthly credit card bill—that same $200 seems smaller, less significant, and less painful than it does on its own. Therefore, when we pay with our credit card, it's easier to undervalue an additional $200 charge. This is a common bias, especially where credit is involved—like spending a few thousand dollars more to upgrade our floors when getting a $400,000 mortgage, or when we easily and without thinking spend $200 more on a car CD changer when we are already spending $25,000 on a new car.

Credit cards are hardly the only financial instrument that embraces the pain-reducing, value-confusing effect of aggregation. Financial advisors make money from investors through various fees. For instance, they generally charge, let's say, 1 percent of our portfolio ("assets under management," as they like to call it). So, as we're making money, they're shaving their fee off the top. We never see that 1 percent. We don't feel its loss because it never reaches our full awareness, so we don't feel the pain of paying it. But what if we paid financial advisors differently? What if every month we had to pay them $800 or so, or at the end of the year we had to cut a check for $10,000 (on our million-dollar portfolio—dare to dream)? Wouldn't that change how we approached their services? Wouldn't we demand much more help? Advice? Time? Wouldn't we look for other options if we were aware of the cost of managing our money?

Or, for those without big investment portfolios, think about all the items in the Smiths' nineteen-page resort tab, or our cell phone bills, where different service purchases and download charges are combined with connection fees. Or cable bundles, where we put phone, Internet, and TV with a monthly subscription to *Bob the Builder* videos because, "Can our toddler figure out the remote?" Yes He Can.

Restricted Access

Let's talk about gift cards again. They're an example of payment tools called "restricted use payment methods," which allow us only to do certain things. Other restricted use payment methods include casino chips and frequent-flyer miles. These make paying remarkably painless. They are already isolated from our normal value cues by mental accounting, but they also make spending easier by removing much of the painful burden of decision-making. If our gift certificate is for Best Buy or our chip only works at Harrah's or our miles are only good on United, then we don't think about whether Best Buy, Harrah's, or United offers the best value. Instead, we mindlessly spend that money there because it's the category that the payment method belongs to, and by doing so mindlessly, we are less likely to critically evaluate our spending decisions.

While we're on the subject of casinos, we may as well point out that they are experts at getting people to part with their money. (The financial industry runs a close second.) From chips to free alcohol, hidden clocks, and twenty-four-hour food and entertainment, they know how to get the most out of every visitor. Remember our friend George Jones from the start of this book, coping with his financial worries at the blackjack table? That's the power of casinos.

There are, of course, countless ways we let the effort of paying affect our spending valuations. The difficulty of paying shouldn't change our sense of value, but it does.

Can You Feel Me Now?

Did you know the first patent that Amazon.com defended was for its "one-click" technology? The ability to buy something—no matter how large or unnecessary—with just one click of a mouse makes spending so easy. So painless. So vital to Amazon.com's success. Online payment, as we've seen, is already incredibly easy. Just a few minutes while we're wasting time on Facebook, and bam! A new sofa is on the way. We're barely even aware that we're spending money.

And that—our lack of awareness of spending—may be the scariest thing about the more and more sophisticated ways with which companies are seducing us into avoiding the pain of paying. So many recent technological advances have made payment so easy that we're often barely aware of our spending. EZ-Pass technology automatically charges us for tolls, and we don't even know the amount until the end of the month (if we bother to check at all). The same is true for automatic bill-pay, where monthly car, mortgage, and other loans are withdrawn without our having to even make one click. Add smart cards, paying by phone, e-wallets, PayPal, Apple Pay, Venmo, probably retinal scans soon enough. These "advances" certainly make paying easier. Frictionless. Painless. *Thoughtless.* If we don't even know something's happening, how can we feel it? How can we understand the consequences? At least in urban legends when villains harvest our kidneys, we wake up in a tub of ice to know something bad happened. Not so with auto-renewing payments.

Salience is the grown-up term for when we're aware of something, in this case, payment. And awareness—having payment be salient—is the only way we could feel the pain, and therefore react, judge, and evaluate the potential costs and benefits of our choices. Feeling the pain is the only way to learn to take our hand off the stovetop.

Paying with cash has built-in salience. We see and feel the money and we have to count it out and then count our change. Checks are slightly less salient, but we do still have to write out an amount and hand something over. As we've discussed, credit cards have even less salience, both physically—just a swipe and the push of a button or two—and in the amount spent. We often barely notice the amount, except perhaps to calculate a tip. Digital payments of all sorts involve even less salience.

If we can't feel it, it can't hurt. Remember, we like things easy. And painless. We'll choose easy and painless over wise and thoughtful every time.

While the pain of paying can make us feel guilty after an expensive dinner, it could also prevent us (to some degree) from impulse shopping. In a future with digital wallets being the main way to pay, there is a risk that almost all friction will be eliminated from the payment system. We are then likely to fall for temptation at a much higher rate. It will be almost as if we spend the whole day lying on a beach full of free drinks, snacks, and desserts within arm's reach. The result? Not good for our long-term health or savings rates.

Our hope is that the future of money will not just be about reducing the pain of paying, but that it will also offer us the opportunity to choose more deliberative, thoughtful, and painful payment methods. With physical money, we have little choice. We have to take the time and attention to pull bills from our wallet and count change. But with electronic money, the temptation is to pick payment methods that hide the pain of paying from us. And if some banks create more painful and deliberate payment methods, will we choose the settings that allow us to feel some of the agony of payment? Will we pick the painful choices that will make us suffer now so that we may benefit later? We should choose a healthy dose of pain now, to remind us that we are spending, to remind us that money neither grows on trees nor on apps. The question is: Will we?

Free Dumb from Pain

What if life were always like Jeff's honeymoon? What if it always felt free? Would we eat more? Enjoy life more in the moment? If something feels free, there isn't any pain of paying, which feels good. But would this actually be good for us in the long term?

Free is a strange price, and yes, it is a price. When something is free, we tend not to apply a cost-benefit analysis to it. That is, we choose something free over something that's not, and that may not always be the best choice.

Say we're going to lunch and we encounter a bunch of food trucks. We're watching our diet and are drawn to a bistro-type vendor that offers sandwiches with a lot of fresh vegetables, with low-fat toppings, on healthy whole wheat bread. Perfect! But then we see another vendor who is celebrating customer appreciation day by offering free deep-fried cheese sandwiches. We've never had any interest in such food, and don't particularly love American cheese, but we're ready to be appreciated. So do we pay for the ideal lunch, or take the not-so-great one for free? If we're like most people, we go for free.

This same type of temptation exists in many parts of our lives, from food to finances. Imagine we have a choice between two credit cards. One offers us a 12 percent APR but has no yearly fee, and another offers us a lower interest rate of 8 percent APR but charges us a $100 annual fee. Most people would overvalue the yearly fee and choose the 12 percent card with no yearly fee. They would end up with a card that costs them much more in the long run, when they inevitably miss a payment or carry a balance. Or let's say we're choosing between two online newspaper subscriptions. One costs $2 a month; the other costs $1.50 a month. In choosing between them, we'll probably consider that one emphasizes foreign coverage, the other political, and decide which interests us more. After all, 50 cents is not much compared with the time we spend reading the newspaper—thus, we can compare the value of the information in each paper. But say the costs are slightly different: What if the first one costs 50 cents and the other is free? Do we still make a careful choice between the two and take into account the value of our time and the value of the content? Or do we simply pick the free, painless option? It's still a 50-cent difference, and reading the newspaper is still an important and time-consuming activity, but when free is an option, most of us would stop thinking, and go for it—all because we want to avoid the pain of paying.

Another effect of free is that once something initially costs us nothing, it becomes very difficult to start paying for it later. Let's face it: When the pain of paying is zero, we often get overly excited—and we get accustomed to that price. Pretend there's an app on our phone that we use to identify songs. We love finding new music, so we listen to college radio stations, check movie soundtracks, and so on. When we hear something we like in a store or in the car, we hit this little app and it identifies the song: Voilà, now we know what this music is! So what happens when one time we try to use this marvelous app and a message pops up informing us that, from now on, if we want to use the app, we must pay a one-time charge of 99 cents? What do we do? Do we pay about a dollar to use something we love? Or do we see if we can find a similar thing for free, even if it doesn't work as well? A dollar clearly is not a lot in the scheme of things, particularly for something that enriches our life. It is not much compared to the amount of money we spend daily on coffee or transit or grooming. And yet, somehow, the change from free to a dollar makes us hesitant to pay for something we've already partaken of for nothing. We don't hesitate to pay $4 every day for a latte, but $1 for an app that used to be free? Outrageous.

Here is an experiment we can all can try: Hold a tray of cups in the middle of a crowded intersection with a sign that says "Free Samples." See how many people take—and ingest—whatever you are offering without even asking who you are, what you are serving, and why. Slightly nefarious, but interesting.

Splitting the Pain

Let's revisit that dinner that Jeff and his wife had with their friends after their honeymoon. There is useful research that suggests that people consume more when everyone knows that the bill will be split, taking some advantage of their unsuspecting dinner partners, as Greg did with the expensive wine.[11] This tendency to over-order when the bill is split evenly suggests that the best payment method is for everyone to pay for what they eat and to declare this strategy at the start of the meal. But is that the most fun? The most pain-free? Far from it.

Taking the pain of paying into account, the recommended method for splitting the bill with friends is credit card roulette. When the server drops off the check at the end of a meal, every one puts down their credit card. The server picks one, and that one person pays the entire bill. A similar, less luck-reliant version of the same thing is to have payment rotate among friends. Everyone takes turns paying the entire bill over the course of several dinners. This method works best if we have a stable group of friends we eat with regularly, though we might be tempted to "accidentally" skip the meal when it's our turn to pay. This last maneuver would help us make fewer payments, but it would also help us have fewer friends.

Why do we like credit card roulette so much? If we consider the utility of everyone at the table—that is, how useful the experience is for everyone around the table, how much enjoyment they get out of it—it is easy to see why one person should pay the entire bill. If every person paid their share, everyone would experience *some* pain of paying. If, on the other hand, just one person paid the entire bill, then the pain of paying would be high for that person, but it would not be as high as the total amount of pain that was saved from everyone else. In fact, it would not be too much higher than if that person just paid for his own meal. The intensity of the pain of paying does not increase linearly with the amount that we pay. We feel badly when we pay for our meal. We do not feel four times more distraught if we pay for ourselves and three friends. In fact, we feel significantly less than four times as badly. And the best feature of this credit card roulette system is that everyone who doesn't pay will eat "pain-free."

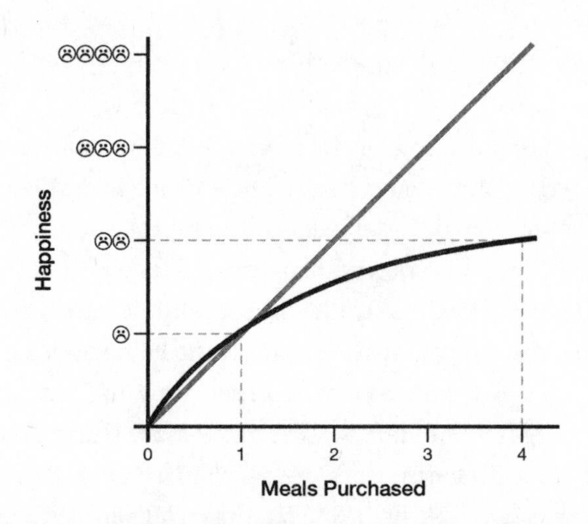

Diminishing sensitivity to the pain of paying for dinner

So, when four people each pay for their meal, we might say the cumulative pain is four frowny faces. When just one person pays, it's just one *very* frowny face and three happy faces. We should also consider the increased collective pleasure from rotating the bill, because our friends get a good feeling when we pay for them, and we, too, feel good about treating our friends to something special.

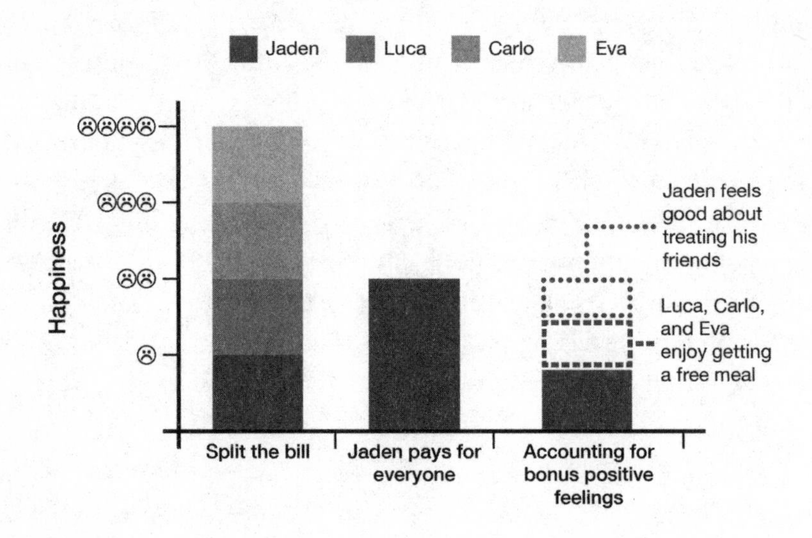

Having one person pay the bill reduces total misery in the long run

This is a classic example of the sports cliché of "taking one for the team," where the team is our friends, and the one is the bill.

Is this system financially efficient? Probably not, because meals cost different amounts, and different people might show up to different dinners and maybe we don't really like some friends as much as others. . . . But even if we end up paying a bit more in the long run for engaging in this practice, we are likely to experience less pain of paying and have more fun dining out. Plus, we will get many more free meals.

The idea of rotating dinner payments shows that the pain of paying isn't, on its own, a bad thing. It's just a thing. Understanding its power can bring some positive benefits to both our financial and our social lives.

We all have pain. We all find different ways to relieve that pain. Some drink or do drugs, some watch *The Real Housewives of New Jersey*, some get married and go on a honeymoon to celebrate a lifetime of having someone else to share in (and maybe blame for) their pain. So long as we're aware of the pain-evading choices we're making, we can help keep them in perspective and limit their impact on our lives.

7

WE TRUST OURSELVES

Way back in 1987, two professors at the University of Arizona—Gregory Northcraft and Margaret Neale—decided to have some fun. They invited some of Tucson's most respected and trusted real estate agents to an open house. These were experts on Tucson real estate, pros who knew the market and the value of a local home better than anyone. Northcraft and Neale allowed the agents to inspect the house and gave them comparable sale prices, information from the multiple listing service (MLS), and other descriptive information.

Each agent got the same information about the house, except for one thing: the price. Some agents were told that the listing price was $119,900. Others were told the listing price was $129,900. A third group was told that the listing price was $139,900 and the last group was told that the listing price was $149,900. (If you own a home in a major metropolitan area today, try not to cry while reading those numbers—it was a long time ago.) The listing price was the first piece of information the agents saw about the house they were checking out.

Northcraft and Neale then asked these expert Tucson real estate agents what they thought was a reasonable purchase price for the home. That is, what was the expected sale price for that home on the Tucson market?

Agents who were told the listing price was $119,900 estimated that the home was worth $111,454. A listing price of $129,900 netted an estimated purchase price of $123,209. A listing price of $139,900 led to $124,653, and $149,900 caused the experts to estimate the value of the house at $127,318.[1]

LISTING PRICE	EXPERT ESTIMATE
$119,900	$111,454
$129,900	$123,209
$139,900	$124,653
$149,900	$127,318

In other words, the higher the listing price—the first price they saw—the higher the estimated price. A $30,000 increase in listing price increased their estimates by about $16,000.

Before we get upset with the ability of these professionals, Northcraft and Neale also tested laypeople using the exact same methods. What they found was that the listing price affected the nonprofessionals much more than it did the real estate agents: The $30,000 increase in listing price caused a $31,000 increase in estimated value. Yes, the professionals were influenced by the initial price, but only about half as much as the nonprofessionals.

But the listing price shouldn't affect a home's value for anyone, in any way, at all. Real estate value should be determined by market conditions like recent sales (comps), by the quality of the home (inspection and MLS info), by the size of the lot, and by the quality of the schools and the competing prices (nearby listings). This should be especially true for experts who know the market and home prices better than anyone, but it wasn't. The listing price clearly affected their value assessments.

Now, here's the most fun part. The vast majority of the real estate agents (81 percent) said they did not consider the listing price *at all* when making their estimates. Of the laypeople, 63 percent claimed they did not consider this information when making their decision. In other words, the listing price changed how everyone valued the property, but most of them had absolutely no idea it was happening.

What's Going On Here?

Who is our most trusted advisor? To whom do we turn for guidance in times of doubt and uncertainty? A parent, a pastor, a teacher, a politician?

It turns out the person we trust the most is—ourselves. That might not be such a good thing. Consciously or not, we rely upon our own brilliance when making value judgments, even though we're not as experienced or as smart as other people and even though we're not as experienced or as smart as we think we are. Our overtrust in ourselves is most pronounced, and most dangerous, when it comes to our first impressions, which is when we are likely to fall prey to anchoring.

ANCHORING occurs when we are drawn to a conclusion by something that should not have any relevance to our decision. It is when we let irrelevant information pollute the decision-making process. Anchoring might not seem too worrisome if we think that numbers don't pollute our decisions very often. But the second, and more dangerous, part of anchoring is that this initial, irrelevant starting point can become the basis for future decisions from that point forward.

The real estate agents in Tucson experienced anchoring. They saw a number, they considered it, and they were influenced by it. They trusted themselves.

When it was suggested that the home should cost $149,900, that number lodged in the head of the agents and became associated with the cost of the house. From that point forward, their future cost estimates had that figure as a reference. It became a personal data point that they trusted, whether they were conscious of it or not.

Just seeing or hearing "$149,900" should have nothing to do with determining the value of a home. It's just a number. But it's not! In the absence of other clear information, in the absence of a verifiable, certain value—and even with a great deal of other context—the real estate experts changed their estimations because they were introduced to that number and from then on were influenced by it. They were drawn in to it like a magnet. Or a black hole. Or, well, an anchor.

Anchors Aweigh

What would we charge to walk someone's dog every day for an hour? How much would we pay for a can of soda? It doesn't take us long to come up with an answer, or at least a range of answers, to these questions. Say we're willing to pay one dollar, at most, for a can of soda. That's our reservation price. Different people generally have a similar reservation price when it comes to something like soda, but why? Do we all like soda to the same degree? Do we all have the same basic level of disposable income? Do we all consider the same alternatives? What processes do we go through to decide how much we'll pay for a soda that makes us all come up with a similar answer?

According to the law of supply and demand, when we set our reservation price we should consider only what the item is worth to us, and our other spending options. In reality, however, we take the selling price into account quite a lot. How much does it usually cost at the grocery store, is it sold at a hotel, or an airport? The selling price is a consideration that is outside the supply-and-demand framework, but like other anchors it ends up influencing the price we are willing to pay. It becomes a cyclical relationship: We're willing to pay about one dollar because that's how much the soda normally costs. This is the effect of anchoring. The world is telling us that the price of a soda is about a dollar, so we pay that price. Once we've purchased a can of soda for a dollar, that decision stays with us and influences how we determine its value from that point forward. We have married a monetary amount with a product, for better or worse, till death—or shaken can of soda—do us part.

Anchoring's impact was originally demonstrated by Amos Tversky and Daniel Kahneman in a 1974 experiment regarding the United Nations.[2] They had a group of college students spin a wheel that, because it was rigged, landed on either 10 or 65. They then asked the students two questions:

1. Is the percentage of African nations in the UN higher or lower than 10 or 65 (whichever number the wheel had landed on)?
2. What is the percentage of African nations in the UN?

For those students whose first question was whether the African nations were higher or lower than 10, the average answer to question 2 was 25 percent. For those who were first exposed to the 65 number, the average answer to question 2 was 45 percent. In other words, the number from the wheel for question 1 made a big difference in the answer to the independent question 2. That first use of the number got them thinking about either 10 or 65 in relation to the percentage of African nations in the UN. Once they'd been exposed to either 10 or 65, that number influenced their own, supposedly unrelated evaluation in the second question. This is anchoring at work.

For those keeping track of obscure and potentially useless information, in the 1970s, 23 percent of the countries in the UN were African.

What this reminds us of is that when we don't know the value of something—how many dollars for a house, how many CD changers for a sunroof, how many African nations in the UN—we're especially susceptible to suggestion, be it from random numbers, intentional manipulation, or the foolishness of our own minds.

As we saw with the pain of paying and relativity, when we're lost in the sea of uncertainty, we cling to whatever object floats by. An anchor price offers us both an easy and familiar starting point.

The Tucson listing price created a starting point for the perception of value, just like the spinning UN wheel. The higher the listing price, the higher the perceived value, even though, as we know, the actual value to us should be based upon what we would pay. What we would pay, in turn, should be based on opportunity cost, not the asking price.

The Tucson story is important because those real estate agents were the most informed and experienced—they were expected to be capable of determining a true value estimation. They were the least lost at sea. If anyone could assess the value of the home in ways that *only* had to do with value, it was them. But they could not. We might say this is proof that real estate is a sham, and, as homeowners, we might agree, but the more relevant point is that if it could happen to these professionals, it could happen to anyone. And it does.

We are all influenced by anchors, all the time, usually without knowing it. After all, remember that 81 percent of the agents and 63 percent of the laypeople said they were *uninfluenced* by the anchor price. The data shows that they were, in fact, *very* influenced, but they didn't even know it was happening.

Anchoring is about trusting ourselves, because once an anchor enters our consciousness and becomes something we accept, we instinctively believe that it must be relevant, informed, and well reasoned. After all, we wouldn't mislead ourselves, right? We can't just be wrong, either, because we're brilliant. We certainly never willingly admit that we're wrong, to ourselves or anyone. Ask anyone who's been in any kind of relationship: Is it easy to admit being wrong? Noooooo. It's one of the hardest things in the world.

The fact that we don't like to admit we're wrong in this case is less about arrogance than laziness (it is not that arrogance is not an important driver of behavior in general; it's just not in this particular case). We don't want to have to make hard choices. We don't want to challenge ourselves when we don't have to, so we go for the easy, familiar decision.

And that decision is often influenced by a starting point anchored into our brain.

Over Herd

Let's think about *HERDING* and *SELF-HERDING* for a moment. Herding is the idea that we will go with the crowd, that we assume something is good or bad based upon *other people*'s behavior. If other people like it, or review it well, or beg to see it, do it, or pay for it, we'll be convinced it's good. We assume something is of high value because others *appear* to value it highly. Herding is essentially the psychology behind review sites like Yelp. It's why we're drawn to restaurants and clubs with long lines outside. Like those giant venues can't let those kids wait inside? No, they want them outside, where they serve as fashionable, attractive beacons herding those seeking to spend their money on designer vodka and booming sounds.

Self-herding is the second, more dangerous part of anchoring. Self-herding is the same fundamental idea as herding, except that we base our decisions not on those of other people, but on similar decisions we ourselves have made in the past. We assume something has high value because *we* valued it highly before. We value something at what it "normally" or has "always" cost, because we trust ourselves with our own behaviors. We remember that we've made a specific value decision over and over, so, without spending the time and energy to evaluate that decision over and over, we assume it was a good one. After all, we are fantastic decision makers, so if we made that decision before, it has to be the best, most well reasoned one. Isn't that obvious? Once we pay $4 for a latte or $50 for an oil change, we're more likely to do so in the future, because we have made this decision before, we remember it, and we're partial to our own decisions—even if it means paying more than we need to. Even if there's a place offering free coffee while we wait for our $25 oil change.

This is how anchoring starts with a single decision, but then grows through self-herding to become a bigger problem, creating a perpetual cycle of self-delusion, fallacy, and incorrect valuation. We purchase a widget at a certain price because of a suggested price—an anchor. Then that purchase price becomes evidence that this was a good decision. From that point on it becomes the starting point for our future purchases of similar widgets.

Another value-manipulating cue that is a close relative of anchoring and self-herding is *CONFIRMATION BIAS*. Confirmation bias pops its head up when we interpret new information in a way that confirms our own preconceptions and expectations. Confirmation bias is also at work when we make new decisions in ways that confirm our previous decisions. When we've made a particular financial decision in the past, we tend to assume that we made the best decision possible. We look for data that supports our opinion, feeling even better about the quality of our decision. As a consequence, our previous decisions are reinforced and we simply follow suit in the present and future.

One need look no further than the way we get our information about the world to realize the power of confirmation bias. We all get to pick the news outlets that we want to give us information, and we do so in a way that rejects information that contradicts our existing beliefs. We focus on news that reinforces and agrees with our preconceived notions. This is not good for us as citizens or as a nation, even if it is a more pleasant experience for us as individuals.

It makes some sense for us to trust our previous decisions: We don't want to spend our lives filled with the stress of self-doubt, and some of our past decisions could in fact be well reasoned and deserve repeating. At the same time, relying on our historic decisions puts a lot of pressure on our past self, on the self who made the first value decision, whether that was the conscious choice to buy a $4 coffee drink or the subconscious choice to consider paying $149,900 for a home. They say we only get one chance to make a first impression. This may be just as true with our financial decisions as with relationships.

Anchoring affects not just real estate pricing, but financial decisions as diverse as salary negotiations (the first offer makes a huge difference in outcomes) to stock prices, jury awards, and our tendency to buy more of the same product when we see a sign that tells us "Buy 12 and get one for free."[3]

There are countless other examples of the effect of anchoring. Will we share more or less than one hundred examples? How many examples do you expect? Ah, now we're just messing with you.

- Let's go back to buying a car. Few people pay the manufacturer's suggested retail price (MSRP), but it is displayed prominently for a reason: anchoring.
- Imagine we're deep in the bowels of a shopping mall, walking by a shoe store. In the window, a pair of glittery pumps beckons to us. What truly catches our eye is the gasp-worthy price tag: $2,500. Two thousand, five hundred dollars for a pair of shoes? We think about this for a few seconds but we are unable to believe it. We walk inside the store anyway and find ourselves holding a different pair of $500 heels that we really, really, really like—but we know we really, really, really shouldn't buy. Oh, but in the land of the $2,500 pump, the $500 shoe is king.
- Prefer food to shoes? Think about sitting at a fancy restaurant looking at a well-designed menu. What do we see first? The luxurious lobster and truffle-encrusted, grass-fed, hand-massaged Kobe beef delight for $125. That's not what we want, or what we get, but it serves to anchor our perspective on the value of other items on the menu, and to make everything else seem affordable by comparison.*
- Executive pay in American corporations has skyrocketed, in part because of anchoring. Once the first $1 million, or $2 million, or

* People like Gregg Rapp, a restaurant consultant, say the highest-priced items actually generate revenue by getting people to buy the *second*-highest-priced items. This is decoy pricing using anchoring and relativity.

$35 million CEO hits the market, that figure raises expectations and estimations about the value of executive leadership—at least in the eyes of other executives. They call this type of pay anchoring "benchmarking," because that sounds better than "screwing people over because they can get away with it."

- Remember the Salvador Assael black pearls from our discussion of relativity? They were placed next to diamonds and other precious gems to make them seem valuable. That placement served to anchor the perceived value of the pearls to our perceived value of diamonds and rare jewels, which, thanks to the efforts of the De Beers family, is quite high.

These and countless other examples show us the many ways anchoring can shift our perception of value.

ZERO ANCHOR

Anchoring can work to keep prices low, too. Just because we save money, that doesn't mean we're valuing things correctly.

Think back to the free apps we discussed earlier. Apps fit neatly into a few price categories, and once these prices have been established, people don't necessarily think about the benefit of the app relative to the benefit they could get from the same amount of money spent on something else. Instead, they focus on the price of the app relative to the initial anchor.

For instance, what if there was a new app that we would use for fifteen minutes, twice a week for a whole year, and it cost $13.50? Is this a low or a high price? It's difficult for people to think about the absolute amount of pleasure and utility they might derive from such an experience compared to other ways in which they can spend their money. Instead, we compare the cost of this app to the cost of other apps, and in the process we deem the new one not worth the money. Wait! This app would give us twenty-seven hours of enjoyment. This is the same amount of time it would take to watch eighteen movies, which would cost around $70 to rent from iTunes and much more than that to see in theaters. This is also equivalent to fifty-four half-hour television episodes, which would cost $53.46 to stream at 99 cents each. When we look at it this way, $13.50 for twenty-seven hours of fun doesn't seem like a bad deal. The problem is that we don't do this exercise—or anything like it. Rather, we compare this app to other apps on price alone—a price that's been anchored to zero. As a consequence, we end up spending our money in ways that don't maximize our pleasure and may not make financial sense.

Ignorance Is Bliss

The less we know about something, the more we depend on anchors. Consider once again our real estate example, where real estate agents and "regular people" in Tucson were shown anchor prices and then asked to assess the value of the home. The real estate experts, who presumably had more than a layperson's understanding of the home's value, were affected less by the anchor prices than were those who didn't know as much. We can also assume that if yet another group were not even given the multiple listing service sheets, comps, and other relevant information, they, with even less knowledge, would be even more swayed by the anchors.

This finding—that anchoring has a weaker effect when we have some rough idea of value versus when we have no idea—is important to keep in mind. When we start with an established value and price range in our minds, it's harder for outsiders to use anchors to influence our valuations.

William Poundstone relays the story of how, after Andy Warhol's death, the artist's property in Montauk, Long Island, went up for sale. Considering the seemingly arbitrary prices of the art world, how could we determine the price of a home that was (sometimes) occupied by a leading art figure? What are the markers for value? His presence, his aura, his fifteen minutes of fame? It was listed for the absurd price of $50 million.[4] Eventually, it was cut to $40 million. If $10 million could have been sliced off the price, why list it for so much money in the first place? Anchoring. The $50 million lingered as an anchor and, soon enough, someone paid $27.5 million. That's about half the original asking price, but, again, the asking price was: Fifty. Million. Dollars. Had the property been originally listed at $9 million—still quite a lot, but closer to the value of area estates—it would have been unlikely to have *risen* threefold. The supersize asking price raised the estate's *perceived* value. It was, perhaps, a fitting posthumous comment on consumer culture by the great painter of brand-name tomato soup cans.

When we encounter a product or service that we can't exactly place, like Warhol's sometimes house, the anchoring effect is powerful. It is even stronger when we are introduced to new products that are simply unlike anything that's come before. Imagine no market, no comparables, no benchmarks, no context for a product or service. For items that seemingly appear from outer space . . .

When Steve Jobs introduced the iPad, no one had ever seen such a thing. He put the figure "$999" on the screen and told everyone that all the experts had said it should cost $999. He talked for a while longer, keeping that price up there, then finally revealed an iPad price of . . . $499! Woo-hoo! What a great value! Heads exploding! Children weeping with joy! Electronic pandemonium!

Dan once did an experiment in which he asked people to report how much they would charge to paint their face blue; smell three pairs of shoes; kill a mouse; sing on a street corner for fifteen minutes; shine three pairs of shoes; deliver fifty newspapers; and walk a dog for an hour. He chose things like smelling shoes and killing a mouse, for which there is no market, so that people could not fall back on familiar techniques to establish their price. For shining shoes, delivering newspapers, and walking dogs, there was a pretty standard price range—around the minimum wage. When people indicated how much they would charge for the activities that had an anchor, they basically came back with a price that was not too different from the minimum wage. But for the first four activities—painting a face, smelling shoes, killing a mouse, and singing—there was no anchor, and the responses were all over the map. Some were willing to do them for almost no money and some wanted thousands of dollars.

Why? When considering something like smelling shoes, we don't know the market price. So we have to start with our own preferences. These are very different for different people, and they're often difficult to figure out. We must dig deep, consider what we like, what we don't like, what we're willing to spend, how much we'd enjoy it, what we're willing to give up (the opportunity cost), and much more. It can be a challenging process, but we have to go through it and eventually we come up with a price. A price that ends up being very different for different people.

When there is a market price for something—like, say, a toaster oven—we don't think through our preferences. We don't have to. We accept the market price as a starting point. We might still think about opportunity costs and about our budget, but we'd be starting from the market price point, not our own, and we'd end up with a final price that is not too far from where we started.

To think about this in a different way, try to express the pleasure of a wonderful, good night's sleep in dollars. Each of us will offer a different answer based on how easily we fall asleep and how much we enjoy sleeping. How much money is that experience worth? It's hard to say. But what if we had to price the pleasure of eating a chocolate bar or drinking a milk shake? We probably know immediately what it is worth to us—not because we just computed the pleasure that we expect from this experience but because we start with the market price and end up very close to it. Similarly, it's hard to determine how much we would have to be paid to allow someone to stomp on our foot for thirty seconds, but if there were a market for getting stomped on, we probably would have an easier time setting our price for that experience. Not because the exercise of figuring out our pleasure is any easier, but because we can use a different strategy (anchoring) to come up with an answer. Not necessarily the right answer—but an answer nevertheless. If nothing else, we hope this inspires some of you to become entrepreneurs in the exciting fields of foot stomping and shoe smelling.

Arbitrary Coherence

As you probably noticed, anchoring can come from both the first price we see, like an MSRP (manufacturer's suggested retail price), and from the prices we've paid in the past, like for a can of soda. The MSRP is an example of an external anchor—that is, the auto manufacturer planting the notion that the car we lust after costs $35,000. The soda price is an internal anchor, coming from our own previous experience buying Coke, Diet Coke, or New Double Diet New Caffeine Free Cherry Coke Zero . . . with Lime. The effects of these two types of anchors on our decisions are basically the same.[5] In fact, not much matters about where the anchor comes from. If we consider buying something at that price, the anchoring effect has been set. The number can even be completely random and arbitrary.

Our favorite anchoring experiments were carried out by Drazen Prelec, George Loewenstein, and Dan. In one of these experiments they asked a group of MIT undergraduate students how much they would pay for certain products, which included things like a computer mouse, a cordless keyboard, some specialty chocolates, and highly rated wines. Before asking the students what price they'd pay, the researchers asked each student to write down the last two digits in their Social Security number—a random figure—and say whether or not they'd buy each item for that amount. For instance, if our last digits were 5 and 4, we would respond whether we would be willing to buy the keyboard for $54, the wine for $54, and so on. Afterward, they asked the students to declare the *real* maximum amount they would pay for each item.

What was so interesting about the results was that the amount the students were willing to pay was correlated to the last two digits of their Social Security number. The higher the number, the more they'd pay. The lower the number, the less. That was true even though—obviously—their Social Security numbers had absolutely nothing to do with the real value of the items, but it did influence the value that they assigned to the item.

Of course, Drazen, George, and Dan asked the students if they thought the last two digits of their Social Security number had any impact on their valuations and bids. They all said no.

This was anchoring in action. More than that, it was completely random anchoring, and yet it influenced the prices. Once even the most random figure is established as a price in our minds, it informs prices for other related products now and in the future.[6] Logically, it shouldn't, but it does. We left logic behind long ago.

That's important and worth repeating: An anchor price can be any figure, no matter how random, so long as we associate it with a decision. That decision gains power and influences our future decisions moving forward. Anchoring shows the importance of early decisions about pricing, that they establish a value in our heads and affect our own value calculations going forward.

This is not the end of the story! Anchors gain their long-term impact with a process called *ARBITRARY COHERENCE*. The basic idea of arbitrary coherence is that, while the amount that participants were willing to pay for any item was largely influenced by the random anchor, once they came up with a price for a product category, that price became the anchor for other items in the same product category. The students in the above experiment were asked to bid on two products within a category—two wines and two computer accessories (a wireless keyboard and a mouse). Did the decision about the first product in a category—the first wine or the keyboard—affect their decision about the second product in the same category? Hopefully it's no longer a surprise to learn that, yes, the first decision influenced the second. The people who first saw the average wine were willing to pay more for the second, better wine. People who saw the nicer wine first were willing to pay less for the second wine. The same was true with the computer accessories.

This means that once we move on from our first decision in a category, we stop thinking about our initial anchor. Instead, we make the second decision relative to the first one. If our Social Security numbers, 7 and 5, randomly get us to pay $60 for a bottle of wine, we price the second bottle of wine relative to the $60 bottle, but independent of the 7 and 5. We are moving from anchoring to relativity. Of course, the anchor still factors in, because it got us to $60 instead of $40, for example, and if we determine that the second bottle is worth half the first, we're spending $30 (half of $60) instead of $20 (half of $40).

In life, we mostly experience relative evaluations. We compare TVs, cars, and homes. What arbitrary coherence shows us is that we can have two rules. We can first determine the baseline price for a category of products in a completely arbitrary manner, but once we make a decision within that category, we make later decisions in that category in a relative way, that is, by comparing them to each other. While this seems sensible, it's not, because starting with an irrelevant anchor means that none of the prices reflect true value.

What Drazen, George, and Dan found was that the random starting points, and the subsequent pattern of valuations that began with these anchors, created an illusion of order. Again, when we don't know what something costs, or when we're uncertain about anything in life, we'll cling to whatever we can. Apps, iPads, no-foam soy lattes, smelling shoes—these aren't, or weren't previously, goods with established prices. Once prices were suggested and we convinced ourselves they were reasonable, the prices became set in our mind, anchored to affect our valuation of similar goods from that point and into the future.

In many ways, initial anchors are some of the most important price markers in our financial lives. They determine a baseline of reality—what we consider real and reasonable for a long time. Most magicians, marketers, and politicians would love to have a trick that is as simple and powerful as the Social Security number anchor. For the rest of us, all these numbers and relativities and prices have made one thing clear: We could all use a drink, of either good or relatively less-good wine.

Raising the Anchor

As teenagers, we often believe that we're invincible. We are superheroes. When we get older, we realize we have limits. We make mistakes. We're not superheroes, we're just people who wear red tights. We realize our physical limitations and the folly of our poor choices. However, we gain insight—sometimes humbling insight, but still—only from decisions about which we're conscious. We don't ever get to doubt decisions that we make unconsciously, that we don't pay attention to, that we've forgotten, or those we've been using thoughtlessly forever as a foundation for our lives.

We really don't know what any particular thing is worth to us. That should be clear by now. That we are so easily and unconsciously swayed by a suggested price—by an anchor—should reinforce how hard it is to assess value. Because it is so difficult, we look for help, and we often turn to ourselves, no matter how wise—or unwise—our past value decisions may have been. We stand on the shoulders of giants . . . even if those giants are the giant mistakes we ourselves have made.

Most investment material includes a disclaimer that says, "Past performance is no guarantee of future results." Considering how much anchoring affects our ability to value items, and how much of anchoring is based upon prior choices, we should apply a similar disclaimer to our lives: Past decisions are no guarantee of future results.

Or, to put the lesson another way: *Don't believe everything you think.*

8

WE OVERVALUE WHAT WE HAVE

Tom and Rachel Bradley are a fictional couple living in Midsized City, USA. They have three kids, two cars, and one dog, and they survive on a diet of wisecracks, sitcoms, and sugary drinks. Rachel is a freelance copywriter and Tom is a senior account manager at WidgeCo, the nation's preeminent producer, distributor, and marketer of high-quality widgets. His job requires him to explain that a widget is merely a term used by economists as a stand-in for a generic good. "Ya see," Tom tells clients about five times a day, "widgets are crucial for your business. They are compatible with your organization and they are the only possible engine of growth. It doesn't matter if you understand what they do, you need to order more now!" He's been there fifteen years.

(For what it's worth, Rachel is named for Jeff's high school crush and Tom is named for his midlife crush, the quarterback of the New England Patriots.)

Tom and Rachel's twins, Robert and Roberta, are off to college, so the Bradleys are downsizing their house. They don't want to leave the area, as their third child, Emily, is just starting high school and has lots of close friends (and some not-so-close frenemies). However, they don't need four bedrooms and they could use the extra money.

They start the process of selling their home by listing it themselves, figuring they could save a commission. They ask for $1.3 million.* Not only do they fail to get any offers, but they also get annoyed. At open houses, potential buyers get distracted by little imperfections. Like some chipped paint, a rusty water heater, "weird" design touches. Tom and Rachel talk about all the great things their kids did in the kitchen and living room, point out where there was a fun scuffle with the dog, highlight all the renovations they've done and the way they designed the layout to maximize space. No one seems impressed. No one seems to see just how great the house is, nor how much of a bargain it is.

The Bradleys finally enlist the help of a real estate agent. Mrs. Heather Buttonedup, the broker, suggests they list it at $1.1 million. They disagree. They both remember their friends selling a similar house down the street for $1.4 million three years ago. They even had a couple of unsolicited offers to buy their place back then, one at $1.3 million and the other at $1.5 million. That was three years ago and now their place must be worth at least that much, if not more, especially considering inflation.

"But that was during a real estate boom," Heather says.

"And it's three years later now, so surely it's increased in value. . . ." pleads Rachel. "And our house is much nicer than theirs."

"Maybe to you, but look at all the work that needs to be done. People don't want an open floor plan these days. The buyer will have to make some real changes."

"What?!" cries Tom. "Do you know how much time, effort, and money we put into making these renovations? It's awesome."

"I'm sure it is to you, but—what is *that*?"

"It's a bike rack."

"Above the kitchen table?"

*Present-day Midsized City, USA, is a very different real estate market from 1987 Tucson, Arizona.

"It adds excitement to every meal."

She rolls her eyes. "Well, it's up to you, but my advice is, if you want to sell this place, list it at one-point-one and be happy if you get close to that."

They'd bought the place fourteen years ago for $400,000, so they'd be making a lot of money no matter what. Still they wonder just how crazy Heather and the potential buyers are if they can't see how special their house is.

After some long nights of deliberations, the Bradleys list their house, through Buttonedup, at $1.15 million. They get an offer for $1.09 million. Heather is ecstatic and says they should take it right away. They want to hold out. After a week, Heather puts on the pressure. "Let's be realistic. Best-case scenario, you wait it out and get another $15,000, $20,000. It's really not worth it. Sell now and move already."

Eventually, they sell it for $1,085,000. The real estate firm of Heather Buttonedup and Associates gets $65,000 on the deal.

Meanwhile, they're looking for a new place themselves. They don't like any of the homes they've seen. They've all had weird redesigns that make no sense and have pictures of kids everywhere. As for the prices, neither Tom nor Rachel can believe the delusion some of these sellers are under, asking way more than their places could possibly be worth. "Do they think it's three years ago when the market was hot?" "Crazy." "Times have changed. Your asking prices should, too."

They finally find a nice house. It's listed at $650,000; they offer $635,000. The seller waits for more. The agent tells them they'd "better hurry and decide quickly because new buyers have emerged." They don't believe her. They end up buying it for $640,00. They're happy enough.

What's Going On Here?

The Bradleys' real estate experience may be fictional, but it is based on many true stories. More important, it shows how we overvalue the things that we own.

In an ideal, rational market, both sellers and buyers should come to the same valuation of an item. That value is a function of the utility and opportunity costs. In most real transactions, however, the owner of an item believes it to be worth more than the buyer. The Bradleys thought that their house was worth more than it was, simply because it was theirs for a while and because they made all these "wonderful" changes to the house—making it even more "theirs." Investing in anything causes us to increase our sense of ownership, and ownership causes us to value things in ways that have little to do with actual value. Ownership of an item, no matter how that ownership came to be, makes us overvalue it. Why? Because of something called the *ENDOWMENT EFFECT.*

The idea that we value what we have more simply because we own it was first demonstrated by Harvard psychologist Ellen Langer and later expanded by Dick Thaler. The basic conceit of the endowment effect is that the current owner of an item overvalues it, and because of that will want to sell it at a price higher than the future owner will be willing to pay for it.[1] After all, the item's potential buyer is not its owner and therefore is not affected by the same love-what-you-have endowment effect. Typically, in experiments testing the endowment effect, selling prices are found to be about twice as high as buying prices.

The price at which the Bradleys wanted to sell their home—how they valued it—was higher than the price buyers were willing to pay. When the roles were reversed and the Bradleys became buyers instead of sellers, the price mismatch also reversed: As buyers, the Bradleys valued the homes they were viewing at lower prices than the owners of those homes valued them.

On its surface, this shouldn't be a surprise. The desire to maximize a selling price and minimize a purchase price is perfectly rational. Basic economic strategy teaches us to try to buy low and sell high. One might assume that this phenomenon is just a simple case of "price high and bid low," right? Not really. This is not a negotiating technique. What careful experiments show is that the higher prices are what owners actually think their possessions are worth and that lower prices are what potential buyers actually think these same things are worth. As we said, when we own something, not only do we start believing that it is worth more, but, furthermore, we believe that other people will naturally see this extra value and be willing to pay for it.

One reason for this overvaluation effect is that ownership gets us to focus more on the positive aspects of what we own.

When the Bradleys were selling their home, they dwelled on good memories—of the spots where Emily first learned to walk and where the twins would fight over who was more loved, of sliding down stairs, of surprise parties, and of all the times they stammered and yelled at their kids using the wrong name. Unintentionally they added those experiences into the joy that the house represented for them and to the value of the home. They simply didn't notice the old boiler or the rickety stairs or the dangerous bike rack as much as potential homebuyers did. They focused on the positives. On the good times.

Even though the Bradleys' reasons for extra value were deeply personal, they were trapped in their own perspective. As a consequence, they expected strangers, without the history of their own experiences, to somehow view the home the same way. Their emotions and memories became part of the unconscious way they valued their home, which of course had nothing to do with the actual value to anyone who did not share in those memories. But when we evaluate our possessions, we are blind to the fact that the emotional boost we get from them is ours and ours alone.

How Do We Own It?

The sense of ownership can and does come in many forms. One of the ways we get an extra feeling of ownership is by investing effort.

Effort gives us the feeling of ownership, the feeling that we've created something. After we invest effort in almost anything, we feel extra love toward that thing we had a part in creating. It doesn't have to be a large part, and it doesn't even have to be a real part, but if we believe we had something to do with the creation, we increase our love and, with that, our willingness to pay. The more work we put into something—a house, a car, a quilt, an open floor plan, a book about money—the more attached to it we become. The more we feel we own it.

The story of effort and ownership doesn't end there. It turns out that the harder it is to make something, the more we feel that we had some part in creating it, and our love for it increases even more.

Mike Norton, Daniel Mochon, and Dan named this phenomenon *THE IKEA EFFECT*—so named after the meatball restaurant/umlaut factory/children's playland that moonlights as a furniture store. Think about what it takes to create a piece of Ikea furniture: We must drive to the massive, rarely convenient Ikea store, navigate the parking lot, watch out for other people's children, grab an oversize bag, follow arrows, look at space-age kitchen equipment, distract our spouse from looking at space-age kitchen equipment, make fun of the names we don't understand, then go pick out our items, lug them to the car, and load them. Then we have to drive home, unload, carry everything upstairs, and spend a few hours swearing at the most pleasant-looking but impossible instructions while convinced that someone must have given us the wrong set of tools because, ah, there it is under my leg, and ow! This doesn't fit quite right, honey, can you just bring up the hammer, yes it's going fine! Done in a few more minutes! I'll just rip that part off, no big deal—it's in the back anyway. Finally, voilà! A nightstand and a lamp! And several extra parts that we quickly hide from our family.

After all that work, don't we feel a strong sense of attachment, a feeling of pride and accomplishment? This is *our* thing; *we* made it! We're sure as heck not going to just toss it aside for a few pennies. That's the Ikea effect.[2]

Think about all the work the Bradleys put into their house. The open floor plan. The pictures. The bike rack chandelier. All of that effort made it feel like something special that they had created. In their eyes, it increased in value with every small change and improvement. The house was such a perfect fit for them and their preferences because of the effort they extended to make it special. Not only did they love their house very much, but they could not believe that others did not fall in love with it the way they had.

We can come to "own" things arbitrarily, without effort. Ziv Carmon and Dan ran an experiment through which they found that Duke University students who'd won basketball tickets in a lottery would only sell them for a price much higher than that which other students (those who did not have a ticket) were willing to pay. That was true even though the ticket was for the same game, the same time, offering the same experience, and the same real value.[3] The lottery winners had no reason to value the tickets more highly than anyone else, except that they owned them. Similarly, other experiments found out that Cornell students who received free mugs valued them at twice the price as those who had no mugs.[4] This isn't just because college students need coffee for anything before 2 p.m., but because those who received mugs randomly very quickly felt that they owned them. Therefore, they overvalued them.

Tangible items are often subject to the endowment effect: People value items more because they have them in their hands. (Perhaps, as we described in chapter 6, this is why AOL used to send CDs with their invitations to use their service, way back in ancient times.) We don't know why mugs are such a popular testing item among social scientists—we'd think red plastic beer cups would be more relevant for college students—but researchers from Ohio State and Illinois State also used them to prove the importance of direct contact. They found that people who held a coffee mug in their hands for more than thirty seconds were willing to pay more to buy that mug than were those who held it for fewer than ten seconds or not at all.[5] Think about that: Thirty seconds is all it takes to establish a sense of higher ownership, strong enough to distort our valuation of an item. That's impressive! Perhaps department stores will mandate that people try on clothes for at least thirty seconds; car dealerships will make us hug a car for a short while; or toddlers will continue to lay claim to every toy they touch by simply yelling, "Mine!"

Consider monthly services that provide free or low-cost trial offers. A magazine publisher offers an introductory rate of $1 a month for three months, a service provider offers a new cell phone that's free for a year, and a cable company offers a cable TV-Internet-phone bundle that's only $99 per month the first year. Eventually, those rates increase—to $20 per month for the magazine, to $30 per month as an add-on to our wireless bill, and another $70 a month to watch shows on TV (shows that we could view on our new phone or read about in our magazine instead).

We could "cancel at any time," but, typically, we don't. Why? Because even though we may not "own" something like cable TV, that trial offer has endowed us with a sense of ownership. Having had and used these services and products, we consider them more valuable, just by virtue of having used them. So when the price increases, it doesn't stop us from continuing the service, because now that we have it, we'll—perhaps begrudgingly—pay more to keep it.

Marketers know that once we possess something—a cable TV package, some furniture, an AOL disk—our perspective will shift. We'll value that good or service more than we would if we had never owned it. Companies employing trial offers are using the same business model as drug dealers: First one's free. Then we're hooked and begging for more. We're not saying cable TV companies are like drug cartels, but we are saying we could stay home and watch most shows online instead (and with our own drug of choice: beer, wine, cigarette, or a pint of Chunky Monkey).

We can also experience something known as *VIRTUAL OWNERSHIP*, which is when we achieve that ownership feeling, enough taste or touch or sense of a product, without buying it completely. Virtual ownership is different from trial offers because we never really own the product.

Imagine we bid on a Mickey Mouse watch on eBay. It's near the end of the auction and we're the highest bidder. We're not the owner yet because the auction isn't over. Nonetheless, we feel like we've won and we're the owner. We start imagining owning and using the product—and are often quite upset if someone swoops in at the last second to outbid us. That's virtual ownership. We never owned it, but it feels like we did, and in the process, we increase how much we value that Mickey Mouse watch.

Dan once spoke to a real estate broker who was involved in a sale of a luxury property, an estate worth tens of millions of dollars. There was a bidding process; negotiations carried on for more than six months. When negotiations began, the bidders had decided what they'd be willing to pay for the property. But as time passed and negotiations dragged on, they found themselves willing to pay more and more. Nothing had changed about the property; there was no new information. Time had simply passed. What had changed? During that time, they began to see themselves as the owners of the property. They thought about how they would use it, how they would live there, and so on. They owned it only in their imaginations—there was no final agreed sale price—but the phenomenon of virtual ownership made them not want to give up the possibility of actually owning it. As the process lingered, their virtual ownership increased and so they valued the estate more and more.

Successful advertising copywriters are, in a way, magicians: They make us feel like we already own their clients' products. We feel like we already drive that car, are on that vacation with our family, or are appearing in photos with those beer-commercial models. It's not real ownership; it's virtual ownership. The fantasies inspired by commercials get us to connect to their product. That connection—the mental touching of the product for thirty seconds—creates a feeling of ownership, which, as we now know, leads to a higher willingness to pay for those products. How long will it be until advertisers use technology to put images of us into the ads we see? That will be us, on the beach, drinking that *cerveza* with those unemployed twenty-year-olds. We just hope they include either virtual weight loss or a virtual appreciation for "Dad bod," too.

It's in the Way That You Lose It

The endowment effect is deeply connected to *LOSS AVERSION*. The principle of loss aversion, first proposed by Daniel Kahneman and Amos Tversky,[6] holds that we value gains and losses differently. We feel the pain of losses more strongly than we do the same magnitude of pleasure. And it's not just a small difference—it's about twice as much. In other words, we feel the pain of losing $10 about twice as strongly as we do the pleasure of winning $10. Or, if we tried to make the emotional impact the same, it would take winning $20 to counteract the feeling of losing $10.

Loss aversion works hand in hand with the endowment effect. We don't want to give up what we own partly because we overvalue it, and we overvalue it partly because we don't want to give it up.

Because of loss aversion, we weigh potential losses much more than we do potential gains. From a cold-blooded economic perspective, this makes no sense—we should consider losses and gains as equal but opposite financial partners. We should let expected utility guide our decisions, and we should just be giant cold-blooded supercomputers—but, thankfully, we're not expected-utility-maximizing machines and we are not cold-blooded supercomputers. We're human (which, of course, is why we'll eventually be ruled by cold-blooded supercomputers).

Owners of an item, like the Bradleys with their home, value the potential loss of ownership much more than nonowners value the potential gain of the same item. This gap—fueled by loss aversion—gets us into all kinds of financial mistakes.

We saw loss aversion at work when the Bradleys referenced the rising and falling real estate market. They thought about the price of their home in terms of its highest point, years ago, before the market slowed down. They thought about what they could have sold it for back then. They focused on the loss relative to the price during that previous historical moment.

Retirement savings and investments are other areas where loss aversion and endowment effect can wreak havoc on our ability to see the world in an objective way. If loss aversion seems like something *we* would never fall prey to, consider your initial reactions to these two questions:

1. Could we live on 80 percent of our current income?
2. Could we give up 20 percent of our current income?

The answers to these two questions should be exactly the same. They are mathematically, economically, and supercomputerly the same question. Can we get by in retirement with 80 percent of our current income? We are, however, much more likely to say yes to question 1 than to question 2.[7] Why? Because question 2 highlights the loss aspect of the situation—losing 20 percent. As we know, losses weigh heavily, so in question 2 we focus on that pain. And what about question 1? That's easy to answer affirmatively, since this question doesn't mention losses at all.

For what it's worth—and it's potentially worth a lot—this same framing issue can arise during end-of-life health-care decisions. When helping families decide whether or not to try heroic measures, medical professionals have found the answer depends on how the decision is framed. People are much more likely to pursue long-shot procedures when they're proposed focusing on the positive—such as "there's a 20 percent change of survival"—than when focused on the negative—like "there's an 80 percent chance of death."[8] May all your loss aversion dilemmas be much less severe.

Loss aversion and the endowment effect can also work together to induce us to turn down free retirement money, like matching funds. Our company might match our retirement contributions, provided we contribute a certain amount ourselves. For instance, if we put aside $1,000, they'll contribute another $1,000, meaning we're getting $1,000 for free. But if we put aside nothing, they contribute nothing. Many people put aside nothing at all; others don't contribute the full amount the company would match. In both cases, they're passing up free money.

Why would we do something as foolish as forgoing free money? There are three reasons. First, contributing to our retirement feels like a loss: We're giving up spending money. We use our salary for many things, like groceries, date nights, wine-of-the-month club memberships. Giving up salary now feels like giving up those things. The second reason is that participating in the stock market creates the possibility of losing money. Voilà: loss aversion (more on that in a moment). Third, skipping the company match doesn't feel like a loss. It feels like passing up on a gain. And, despite how logical we all might feel when calmly reasoning that there's little difference between a "loss" and an "unrealized gain," that's not how we act or how we feel. Don't believe us? Read on for proof.

In one experiment Dan conducted, people were asked to imagine that their annual salary was $60,000 and that their employer would match their retirement contributions, up to 10 percent of that salary. Participants were given expenses like food, entertainment, and education. They had to make trade-offs, as we all do, because the $60,000 was not enough for everything in this experiment—such is life. Few people maxed out their retirement contributions and most people put little away at all. Thus they didn't get the full matching funds.

In a slight variation of that experiment, researchers told another group of participants that their employer had put $500 monthly into their retirement account at the start of each month. Employees could keep as much as they wanted, but to do so they would have to match that amount by making their own contributions. For instance, if they also contributed $500 a month into their account, they'd keep the entire pot. But if they only saved $100, they'd keep only $100 of their employer's contribution and the other $400 would disappear from their account and go back to the employer. Every month, participants who didn't fully fund their retirement accounts received reminders that they had lost the unmatched free money. They were told how much the company prefunded in the account, how much the employee contributed, and how much money the company took back. The statement might say, "We prefunded the account with $500, you contributed $100, and the company took back $400." That made the loss very clear. It also triggered loss aversion in participants, who quickly began maximizing their 401(k) contributions.

Once we understand loss aversion and that many things can be framed as either gains or losses—and that the loss framework is more motivating—maybe we can reframe choices, such as how much to contribute to retirement savings, in a way that will persuade us to act in ways that are more consistent with our long-term well-being.

Speaking of long-term well-being, loss aversion also clouds our ability to gauge long-term risks. This problem specifically impacts investment planning. When risk is involved and the amount of our investment fluctuates up and down, we have a hard time seeing beyond our potential immediate losses to imagine future gains. Over the long term, stocks outperform bonds by a large margin. But when we just look at the short term, there will be many short periods with painful losses.

Let's imagine stock prices go up 55 percent of the time and down 45 percent of the time. That's pretty good. But it's also over the long term, not just a few weeks, months, or even a year.

The trouble is that we experience the up-and-down periods quite differently. During the ups, we are a little bit happy, but during the downs, we are miserable. (As we said earlier—if we can quantify happiness—we're about twice as miserable in the downs as we are happy in identical ups.) By weighting more heavily the down market's impact on us, we don't feel the overall trend as 55 percent up and happy, but as 90 percent down and unhappy (45 percent times two).

Because of loss aversion, when we look at investing in the stock market in the short term, we suffer. In contrast, if we could only view the stock market with a long-term view it would feel much better to take more risks. In fact, Shlomo Benartzi and Dick Thaler found that

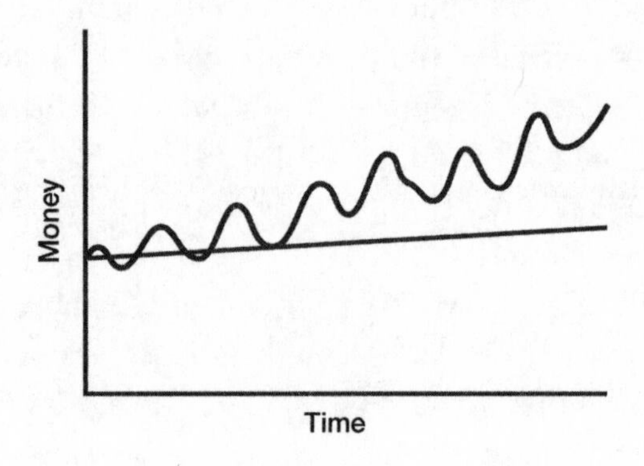

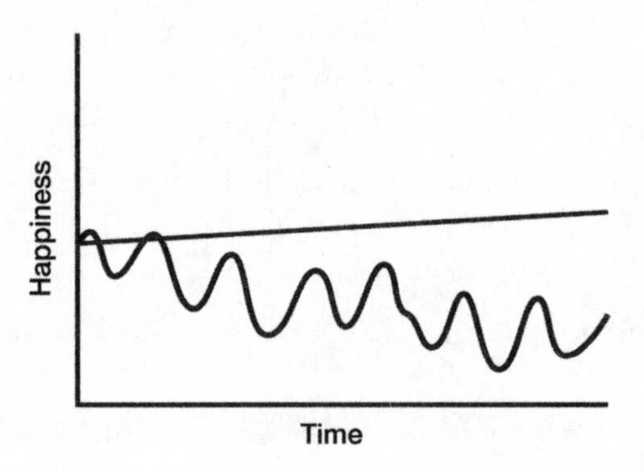

The dark line represents a fixed interest rate, while the gray line represents fluctuating returns. The top graph represents the amount of money involved, while the bottom graph represents the psychological reaction to these gains and losses, taking into account loss aversion such that losses are twice as impactful. Note that while the absolute amount of money is greater in the fluctuating returns case (top graph), as an experience it is more negative.

employees are willing to invest more of their retirement savings in stocks if they are shown long-term rates of return rather than short-term ones, because when we see the long-term view, loss aversion isn't in play.[9]

Loss aversion can create a myriad of other investment problems. In general, it gets us to sell winning stocks too quickly—we don't want to lose those gains!—and keep losing ones too long—because we don't want to realize the loss on those stocks.[10]

One solution people use to avoid the pain of short-term loss is to avoid scary, risky stocks and invest in bonds in the first place or sometimes in saving accounts that give us a certain, but close to zero, interest rate. Bonds don't have the same downs—or ups—as stocks. We don't suffer the loss aversion and we're not as miserable. Of course, we can become miserable in other ways since we reduce our potential for long-term growth. But we don't feel that loss in the moment. We only feel it at retirement, when, sadly, it's too late to change our mind and our investment decisions.

Another approach that we—Dan and Jeff—prefer is to simply not look at our investments. If we're very sensitive to small fluctuations over time, one solution is to simply make a long-term decision and stick to it. Then we don't let loss aversion influence us to act rashly. We (try to) look at our portfolio only once a year. In short, we recognize our irrationality, and we know we are not going to win in a direct fight against it, so we try to avoid the battle altogether. It's not exactly Sun Tzu's *The Art of War*, but we recommend this approach to you as well.

But Wait! There's More!

Ever notice that many companies charge a single amount for what they pitch as many items? For instance, cell phone companies charge us for every little thing that we do—texts, calls, data, FCC charges, equipment rental, line fees, etc.—but in their kindness and their desire to help us not feel several small losses, they ask us to make only one larger payment. What a great deal! We feel one loss, but gain many valuable things.

The cell phone approach is known as aggregating losses and *SEGREGATING GAINS* and it plays on loss aversion, giving us just one painful loss against many pleasurable gains. When a product has many features, it's in the seller's interest to highlight each one separately and to ask one price for all of them. To the consumer, this promotional practice makes the whole seem much more appealing than the sum of its parts.

Depending upon one's religious inclinations, one might imagine God holding court with some angels, reflecting upon the story of Creation. "Yeah, I know how segregating gains works. Like it really took me a week to create the earth with all those things! Ha! Light, fish, animals, trees. It's just one world! One thing. But hey, if humans want to think of it as taking six days, each with a handful of creations, that's fine by me. I'll even set aside a seventh day for rest and football."

The best examples of segregating gains are probably infomercials. The Sham Wow, the Ginsu knife, the ten-CD collection of the greatest big-hair rock songs of the eighties—all of these infomercials present one low price for multiple items that have multiple uses and comes with multiple add-ons. "It's got a top! And a bottom! And not one, but *two* sides! Order now!"

This is why, when Jeff proposed to his wife, he considered acting like an infomercial. "If you say yes now, you'll not only get my hand in marriage, you'll get my arm, and another hand, and another arm. . . . A torso, a head, a wardrobe, some student loans, a Jewish mother-in-law, and so much more! Act now, and we'll throw in not one, not two, but six nephews and nieces! You'll be buying birthday presents year-round! But hurry, this offer won't last long. Our operators are kneeling by, so say yes now!" He almost did that, because he likes a good story, but he was worried about the potential loss from such a proposal, so he went with the less risky, traditional "Will you marry me, pretty pretty please" approach. It worked. *Phew.*

You Sunk My Ownership

Our tendency to emphasize losses over gains and to overvalue what we have plays out very powerfully with *SUNK COSTS*.

Sunk cost is finding that once we've invested in something, we have a hard time giving up on that investment. Thus we are likely to continue investing in the same thing. In other words, we don't want to lose that investment, so often we throw good money after bad, adding a dash of wishful thinking. What if we were the CEO of a car company and we have a plan for a new car that will cost $100 million to develop? We've already invested 90 of the $100 million needed, and all of a sudden we learn that our competitor is nearly finished with a car that's greener, more efficient, and more affordable. The question is, do we abandon our plan and save the last $10 million, or do we spend the last $10 million, hoping that someone will buy our cars despite their inferiority?

Now, imagine the same situation, only this time we haven't invested the first dollar and the total expected cost of development is just $10 million. Just as we plan to start really working on this project, we hear that the competitor has designed a car better than ours. Do we invest the $10 million now? At this decision point—the question of whether to invest $10 million or not—these two cases are exactly the same. However, in the first case, it's difficult not to look backward and see the $90 million we've already spent. In the first kind of situation, most people keep investing. In the second case, they don't even come close to putting in any money. The rational person would make the same decision in both instances, but few people do. The metaphor for investing in many things in life should be the same: We shouldn't think about how much we have already invested in a job, a career, a relationship, a home, or a stock; we should focus on how likely it is to be valuable in the future. But we're not that rational, and it's not that easy.

Sunk costs are costs that are permanently in the loss column of our life-ledger. They are ours, we can never get rid of them, we own them. We don't just see the dollar amount, we see all the choices and efforts and hopes and dreams that went along with those dollars. They become weightier. And since we overvalue these sunk costs, we're less willing to give them up and we are more likely to dig ourselves deeper into a hole.

One way Dan demonstrates to his students the concept of sunk cost is through a game in which participants bid to purchase a $100 bill. Rule #1: Bidding starts at $5. Rule #2: Bids can only increase by $5 at a time. Rule #3: The winner pays the amount of his or her final bid and gets the $100. The last rule is that the second-highest bidder also pays what he or she has bid, but gets nothing. As the game progresses, the bids rise to $50 and $55, at which point Dan will have made money. (The $55 bidder will pay $55 to get $100 and the second bidder will pay $50 and get nothing.) At some point, someone bids $85 and a competitor bids $90. At that point, Dan stops them and reminds them that the first person will win $10 ($100 minus $90) and the second person will lose $85. He asks the $85 bidder whether they want to continue to $95. Inevitably, they say yes. Then he asks the first person the same question, and he happily agrees to go to $100.

But it doesn't stop there at $100. Next, Dan asks the person who's bid $95 if they want to go to $105. As before, if they say no, they'll lose their previous bid: $95. But at this point, when the bidding is over $100, if they say yes, that means they are now actively bidding knowing that they will lose money. This time it's $5 ($105 bid minus $100 winnings), but the loss will only increase from there. Inevitably, both participants keep bidding higher and higher until at some point one person realizes how crazy this is and they stop (and the person stopping ends up losing $95 more).

As Dan tells it: "The most I've ever made off this game was in Spain, where I once sold a 100-euro bill for 590 euros. To be fair, I always tell people up front that the game is for real, and I always take their money in the end. I figure they're more likely to learn their lesson that way, and moreover, I have to keep my reputation."

In Dan's game/experiment/scam, the effect of sunk costs quickly turned his students'/subjects'/marks' potential 95-euro gain (100 euro minus the 5-euro starting bid) into a 490-euro loss. This is just like a contest between two companies in a winner-takes-all market. In general, one company will get all the sales or at least the vast majority, and the other will get nothing. Every quarter, each company must decide whether to invest more in research and development and advertising or to give up the competitive project. At some point, it should be clear that if the two companies perpetually try to outbid each other, they'll both end up losing lots of money. Regardless, because it's hard to ignore past investments, it's difficult not to keep going. The trick to this type of market competition (and the key to Dan's game) is either never to play in the first place or, if we play, to learn quickly when things are not going our way and cut our losses.

Hal Arkes and Catherine Blumer showed one other way in which we don't think clearly about sunk costs. They asked people to assume they had spent $100 on a ski trip (it was 1985). Then they presented a ski trip that was better in every way but cost only $50, and they asked the participants to imagine they bought that one, too. Next, Arkes and Blumer told the participants that the two trips overlapped but there were no refunds available. Which trip did they choose, the $100 okay vacation, or the much better one that was only $50? More than half the participants chose to go on the more expensive trip, even though 1) it was inferior in terms of the pleasure it would provide and 2) they'd spent $150 total either way.[11]

Sunk cost applies to decisions in our personal lives, too. A friend of Dan's was conflicted about whether to get divorced. His life was consumed by this decision. At some point, Dan asked him a simple question: "Imagine that right now you were not married to this person, and you knew about her everything you now know, but you've just been friends for the last ten years. Would you now propose to her?" The friend said there was zero percent chance he would propose. At that point, Dan asked, "What does this tell you about your decision?" How much of his conflict came from thinking about the past, from overvaluing the time and energy he'd already sunk into his marriage, rather than looking forward, to the time and energy he'd use in the future, regardless of the previous investment? Once Dan's friend understood this perspective, he quickly decided to divorce. If anyone thinks this is a heartless way of making a decision, we would like to add that the couple didn't have children, and sometimes giving up sunk cost and looking at things with fresh eyes is good for everyone.

The point is that in many aspects of life, the existence of a past investment doesn't mean we should continue on the same path; in fact, in a rational world, the prior investment is irrelevant. (And if the prior investment has failed, that's a "sunk cost"—we've spent it no matter whether it's failed or succeeded. It's gone.) What is more relevant is our prediction of value in the future. Sometimes looking just at the future is the right thing to do.

Own the Future

Ownership changes our perspective. We adjust to our level of ownership and it becomes the baseline by which we judge gains and losses.

One way to overcome the traps of ownership is to try to separate ourselves psychologically from the things that we own, in order to more accurately assess their value. We should think about where we are now and what will happen going forward, not where we came from. This is, of course, much easier said than done, especially when we tend to put so much emotion, time, and money into our lives and into our possessions—our homes, our investments, and our relationships.

Ownership made the Bradleys focus on what they were losing—their beautiful, personalized house—rather than on what they were gaining for the future—money to buy another house, have some nice dinners, and pay for Robert's and Roberta's tuitions at a good college that is close, but not too close. About ninety minutes is the right travel time to enable Tom and Rachel to visit regularly, but it's not so close that they'll end up doing their kids' laundry every week. They'll miss their kids, but not *that* much.

9

WE WORRY ABOUT FAIRNESS AND EFFORT

It's early morning and James Nolan is in a meeting. Well, it's a presentation. It's probably a waste of time, but it's part of the job. The widget company for which he works (widgets are having a moment) had him hire an outside consulting firm to identify and address deficiencies in their operation. After six weeks, James and his fellow mid-senior-level executives are seeing the results. That is, the results are being shown to them with many PowerPoint presentations.

Gina Williams, the consultant's project head, struggles into the conference room carrying three large binders. She drops them on the table with a thud. Then four junior consultants, two assistants, a tech guy, and a security guard carry in some AV equipment, more binders, a projector, reams of paper, a tub of coffee, and a tray of pastries. James isn't sure why they didn't set up before the meeting, but sugar and caffeine are his gateway drugs into not caring too much, so he eases into his chair and lets the day unfold.

The consultant's team sets up. Then Gina meticulously plods through a seventy-four-slide PowerPoint spiel, detailing everything from the time they boarded their flights two months ago, to all the meetings, and paperwork, and locations, and other meetings, and meals, and supplies, and there are lots of arrows and acronyms. There's a twenty-minute break, then a few slides of credentials and pictures of Gina's family and call logs. It's a five-hour presentation. The last slide—the conclusion—says, "Ask not what your widget can do for you, but what you can do for your widget?"

Everyone in the conference room spontaneously leaps with excitement into a standing ovation. Pastry crumbs fall to the floor, hearty handshakes await by the door, and out into the fluorescent hallway the consultants go, marching toward the future with a newfound sense of achievement and purpose. Huzzah!

Later that day James passes by the executive office suite and observes his CEO gladly cutting a $725,000 check for the project. An inapplicable, repurposed JFK quote for $725,000? Considering all the work they did, it totally seems worth it.

James leaves work early that afternoon to get a $50 oil change. He drops his car off with the mechanics at their otherwise empty shop. They look up from their card game and say it'll be a few hours.

Feeling a little spry after seeing the project with the consultant through, he decides to take the two-mile walk home. Unfortunately, when he's about halfway, the skies open up and an unexpected downpour drenches him. He hustles to a local convenience store to take shelter, and he notices the owner pulling a rack of umbrellas out from behind the counter. James heads over to pick one out but stops when he sees the owner take off the "$5" sign and add a handwritten "$10" one in its place.

"What are you doing? Those are $5."

"Nope, $10. Rainy-day special."

"What? That's not special. That's robbery!"

"You are more than welcome to shop around for a better deal." The store owner motions outside, where it's water as far as the eye can see.

"That's ridiculous! You know me. I come here all the time."

"You should buy an umbrella next time. Every now and then, they are on sale for just $5."

After rolling his eyes for a few seconds, James mutters something unprintable, pulls his collar up over his head, and runs outside umbrellaless, around the corner of the building and all the way back home. As soon as he gets home and peels off his soaking clothes, the rain stops. Another unprintable outburst before scampering half-naked up the stairs.

The auto shop calls to tell James his car required more work than they thought; they're going to have to keep it overnight. They hang up before he can protest. Frustrated, James decides to go outside again for a jog, to burn off some angst. When he finishes, he realizes he's locked himself out of the house. Ugh. His wife, Renee, hasn't yet returned from a business trip. The kids are at friends' houses, and his neighbor with the spare key is on vacation. And it looks like it will start to rain again soon. Reluctantly, James calls a locksmith. Then he calls two more. Each one says it will probably cost $150 to $250 to come out and either pick the lock or replace it entirely. He was hoping for less, but when he realizes that all of them are robbers disguised as locksmiths, he books the last guy. Twenty minutes later, the locksmith gets to the house, approaches the door, twists a thingamabob, jiggles a doohickey, yanks on a whatsit, and, voilà, the door opens. Took him about two minutes.

They go into the kitchen for a glass of water, and the locksmith says, "Thanks. That'll be $200."

"Two hundred dollars? That took like a minute! So you mean that your rate is"—fidgeting with his fingers—"$12,000 per hour?!"

"I don't know about that, but you owe me $200. Or we can go outside and I'll lock you out again and you can try your luck with someone else. It'll take about a minute. Up to you."

"Fine." James writes him a check and slinks down the hall to put on Netflix and enjoy a few minutes alone in the house.

Renee gets home later that evening in a great mood. Her trip was a success and she was happy about having used Kayak—the airfare search service—for the first time and gotten what seemed like a great deal. She'd taken an Uber back from the airport, since the car was in the shop. Renee loves Uber. More than an Uber fanatic, an uber-Uber fanatic. She has an unpredictable schedule, so using Uber saves a lot of hassle scheduling car availability or figuring out public transportation.

A few days later, while her Uber-dom is unblemished, there's a snowstorm the day she needs to go to dinner with a client. It's hard to get an Uber. The normal $12 ride downtown now costs $40. Forty bucks! Outrageous! She calls a regular car service and decides to stop using Uber in protest. Over the next few weeks, she books her old car service, rides the bus, borrows the car, and makes do otherwise. It's a pain, but she's doesn't like being ripped off.

What's Going On Here?

This is how *FAIRNESS* impacts our perception of value. Most people above age five, and not actively engaged in politics, understand the concept of fairness. We recognize it instantly when we see it or talk about it, but we don't realize how great a role fairness plays in our everyday money decisions.

The value provided to us by a consultant's advice, an umbrella in the rain, an unlocked door, or a ride home ought to have nothing to do with whether we *think* the price is fair. And yet, whether we buy something or not, the amount we're willing to pay for things often depends, to a large degree, on how fair the price appears to be.

When evaluating a transaction, traditional economic models simply compare the value to the price. Real, human people, however, compare value to price plus other elements, like fairness. People can actually resent the efficient, perfect economic solution when it feels unfair. That feeling affects us even when a transaction makes sense, even when we would still get a great value—like paying more for a device that would get us home dry.

By the basic laws of supply and demand, umbrellas should cost more in the rain (more demand) and Uber rides should cost more in a snowstorm (lower supply and more demand) and we should be perfectly okay paying these higher prices. The value of getting an oil change or an unlocked door should have nothing to do with a sense of fairness, just that it gets done quickly and efficiently. Still, we fret, roll our eyes, stomp our feet, kick the dirt, and threaten to take our ball and go home when we pay a high price for something that looks easy or takes little time. Why? Because we are little brats who believe that prices should be fair. We will refuse good value because we believe it is unfair. We punish unfairness, and often ourselves (witness James, our soaking-wet widget executive), in the process.

There is a well-known experiment that shows the ways in which we punish unfairness. It's called the ultimatum game. Despite the suspense movie sound of the name, it does not involve Jason Bourne.

The basic setup involves two participants—a sender and a receiver. The two players sit in different rooms. They don't know each other and will never meet this way. They can act in any way they want without fearing retaliation from the other person. The sender is given some money—say, $10. He or she then decides how much of that cash to give to the receiver, while keeping the rest for him- or herself. The sender can give any amount—$5, $1, $3.26. If the receiver accepts the offered amount, they both get their allotted cash, the game is over, and they each go home. If the receiver rejects the offered amount, neither participant gets anything and the money goes back to the experimenter. Nada. Zilch. Zero-point-zero.

Both parties understand the rules of the game, like the amount of money in question and how the money is being split, or not.

If we step back and think rationally, logically, cold-blooded supercomputer-meets-Jason-Bourne-y about it, we'd conclude that the receiver should accept any amount from the sender that is above zero. Even a penny is something they've gotten just for showing up. It's free money, and any sum should be better than getting nothing. If the world were super-rational, the sender would offer one penny and the receiver would accept it. Game over.

But that's not what real people do in the ultimatum game. Receivers routinely reject offers that they consider unfair. When the sender offers less than a third of the total amount, the receiver most often rejects the offer and they both go home with nothing. People actually refuse free money in order to punish someone—someone they don't know and probably won't deal with ever again—just for making an unfair offer. These results show that we can value a dollar at less than zero because of our sense of fairness.

Think about it: If we were walking down the street and strangers were handing us $50 bills, would we refuse them because they were keeping $100 bills for themselves, or would we thank them and remind ourselves to walk down that street every day for the rest of our lives? If we were running a marathon and someone handed us a cup of water, would we toss it aside because there was a table full of cups we were not getting? No, that would be insane. Why is it, then, that in so many other cases, we focus on the half-empty part of the glass—the part that is not fair? The part we are not getting?

Well, maybe we are insane. Researchers found that unfair offers in the ultimatum game—like $1 out of $10—activate different regions of the brain than do fair ones—like $5 out of $10. Research shows that once our "unfair" regions are activated, we are more likely to reject unfair offers.[1] In other words, our brains don't like unfairness and this dislike makes us take action to express our displeasure. Stupid, crazy brains. We may not like them, but they are our brains.

PLAYING WITH ECONOMISTS

The exception to the rule that we reject unfair ultimatum game offers is that economists do not reject unfair offers. They recognize the rational response. Since this is clearly a passive-aggressive attempt to demonstrate how much smarter they are than the rest of us, if we ever play the ultimatum game with an economist, we should feel free to be as cruel and unfair as we want. After all, they have been trained to see low offers as the desired rational response.

James rejected an unfair umbrella price, even though he needed it, he could afford it, and $10 was probably a good value at that time for helping him stay dry. James didn't reject the locksmith's work, though he clearly expressed displeasure and frustration, undervaluing quick access to his own home. Renee quit Uber for a while after experiencing Uber's weather-related price hikes, even though the value of using this service under regular weather conditions remained the same.

(For those paying close attention, yes, James refused to spend an extra $5 to stay dry *on the same day* on which he didn't flinch at his boss paying $725,000 for a long-winded PowerPoint presentation. There's a reason James's brain didn't perceive these two transactions as contradictory. Hang on, we'll get to it soon.)

What if Coke machines were equipped with thermometers and were programmed to charge more money the hotter it got outside? How would we feel about this on a 95-degree day? This was a suggestion made by Douglas Ivester, chief executive of the Coca-Cola Company, to raise revenue. After consumers reacted with outrage, and Pepsi called Coca-Cola an opportunist, Ivester was forced to resign—even though the company never produced any such machine. The supply-demand pricing strategy was logical, perhaps even rational, but people perceived the idea as unfair. It seemed like a barefaced attempt to gouge customers, and boy, it made people angry.

We certainly appear to have a dormant "harrumph" lurking in our economic dealings. We like to tell our trading partners, "Do not profit at my expense!" We are grumpy, judgmental people: We pass up good value that seems unfair, out of spite and in search of revenge.

When our sense of fairness is engaged, we don't care if there are legitimate reasons for a higher price. The invisible hand of the market gets smacked away. In a telephone survey (remember telephones?), 82 percent of respondents said that it was unfair to raise shovel prices after a snowstorm (a hybrid of umbrellas in the rain and Uber in the snow), even though the standard economic rule of supply and demand makes it the efficient, legitimate, correct thing to do.[2]

In 2011, Netflix announced, in a blog post, that it would soon change its pricing structure. It would split its combined streaming and DVD rental services, at the time costing $9.99 per month, into two separate services, each of which would cost $7.99 per month. So, if we primarily used one service—streaming or DVD rentals—our price would drop by $2 per month. But if we used both, the total price would rise by almost $6 per month.

Most Netflix subscribers used only one of its services, but what do you think their reaction was to the change? Yup. They hated it. Not because the price was worse—in the vast majority of cases, it was better—but because it seemed unfair.* These loyal Netflix-loving customers went all JCPenney on Netflix's derriere. The company lost about a million customers and its stock price tanked. Within weeks, Netflix execs scrapped their new plan. Because people felt that Netflix was profiting at their expense, they rejected a service that still had a tremendous value to them—a value of at least $9.99 for which they'd only have to pay $7.99. Netflix customers wanted to punish the unfairness, and they were willing to hurt themselves financially by doing so. They were willing to forgo a wonderful service that was now $2 cheaper, just to punish the imaginary $6 increase of the combined services they didn't even use.

Renee's experience with Uber is based on a true story (as are all the cases we discuss here). In December 2013, during a snowstorm in New York City, Uber charged prices up to eight times its normal rate—a rate that was already higher than regular taxi and car services.[3] Celebrities

* Also at play here is loss aversion—customers didn't want to give up their DVD option, even if they didn't use it.

were among those most vocal about their outrage (they have time to be outraged). Uber responded that the new rates were simply "surge pricing": a spike in fares to lure more drivers onto unsafe roads. But it didn't calm people down.

Uber's customers *normally* enjoy the reliability and availability of Uber's drivers and are willing to pay some premium for that availability. But when true market forces of supply and demand come into play in a big way, as in a snowstorm, when driver supply is down *and* demand is up, thus raising prices a lot, customers suddenly balk at paying the premium. If there were no Uber, there wouldn't be enough taxis, and riders would have little chance of getting one. Uber charges extra to fight such imbalances between when riders need rides and when drivers want to offer them. On a regular basis we're willing to alter our perception of fair price and fair value—but only a little bit. Our flexibility has a breaking point. When a premium is large, sudden, and opportunistic, it feels unfair.

As a further thought experiment, imagine there was a different car service, called Rebu. It always charged eight times more than Uber. In that case, customers would have been fine paying Rebu's higher prices during the snowstorm. That's Rebu's normal rate. In fact, they might have considered this a deal. It was only because Uber raised its rate right when people needed transportation the most that they thought it unfair. If Rebu's rate was always eight times Uber's, it wouldn't have seemed unfair during the snowstorm—though it might have seemed overly expensive every other time.

Fair Effort

Why does the principle of fairness change our perception of value? Why do we discount things that we believe are unfair? Why did Renee abandon Uber and why did James run through the rain? Because fairness is deeply rooted in us. And what makes us see things as fair and unfair? It is largely about effort.

Assessing the level of effort that went into anything is a common shortcut we use to assess the fairness of the price we're asked to pay.

Selling umbrellas doesn't get harder because it's raining. Driving for Uber during a snowstorm might require a little extra effort, but not eight times as much. These price increases don't seem to match the extra effort, and without any increase in the cost of production, we believe that the price hike is unfair. But what James and Renee miss when they focus only on effort (and, thus, fairness) is that the value of the service to them—getting home safe and dry—has *increased* because of the new circumstance, even if the effort required by the service provider didn't change.

James didn't think the locksmith's price was fair, because it took him so little time. But would he have preferred that the locksmith bumble around, take a long time, and fake effort? Well, maybe. A locksmith once told Dan that when he started his career, he took forever to open a lock, and in the process, he often broke it, taking even more time and money to get one properly installed and finish the job. He charged for the parts to replace the broken lock as well as his standard fee for opening a locked door. People were happy to pay all this, and they tipped him well. He noticed, however, that as he became proficient and opened a lock quickly, without breaking the old lock (and without the consequent need to replace it and charge his clients for the extra parts), customers not only didn't tip, but they also argued about his fee.

Wait, what? How much is it worth to have our door open? That should be the question. But because it's difficult to put a price on this, we look at how much effort it takes to have that door unlocked. When there's a great deal of effort, we feel much better about paying more. But all that should matter is the value of that open door.

That's how our unconscious blending of effort and value often leads us to pay more for incompetence. It's easy to pay for conspicuous effort. It's harder to pay for someone who is really good at what they are doing—someone who performs the job effortlessly, because their expertise allows them to be efficient. It's hard to pay more for the speedy but highly skilled person, simply because there's less effort being shown, less effort being observed, less effort being valued.

On Amir and Dan once did a study in which they asked people how much they would pay for data recovery.[4] They found that people would pay a little more for a greater quantity of rescued data, but what they were most sensitive to was the number of hours the technician worked. When the data recovery took only a few minutes, willingness to pay was low, but when it took more than a week to recover the same amount of data, people were willing to pay much more. Think about it: They were willing to pay more for the slower service with the same outcome. Fundamentally, when we value effort over outcome, we're paying for incompetence. Although it is actually irrational, we *feel* more rational, and more comfortable, paying for incompetence.

There is a legend that Pablo Picasso was approached in the park by a woman who insisted he paint her portrait. He looked her over for a moment, then, with a single stroke, drew her a perfect portrait.

"You captured my essence with one stroke. Amazing! How much do I owe you?"

"Five thousand dollars," Picasso replied.

"What? How could you want so much? It only took you a few seconds!"

"No, ma'am. It took me my entire life and a few more seconds."

This is where expertise, knowledge, and experience matter, but these are also the exact same things we fail to value, we lose sight of, when we make value judgments based primarily on effort.

Here's another scenario. Ever had a stubborn car problem—say, a noise or a window that won't budge—and the mechanic fixes it in a few minutes with one simple tool and turns around and tells us that will cost $80? Most people get angry in that circumstance. Now consider if it had taken three hours and cost $120. Would that seem more justified? What if it took four days and cost $225? Isn't the problem fixed either way, and at a fraction of the time and cost in the first scenario?

Think about a computer repair technician, who can fix our company's vital server by changing one configuration file. Our company is paying not just for the simple change—a five-second effort—but for knowing which file to change and how. Or what if we are trapped with an action movie hero who's trying to defuse a nuclear bomb. The seconds are ticking down to zero. The fate of the world is at stake—all will be lost! Would we rather he fumble around, poking and prodding the explosive device with clumsy fingers, or would we pay a fortune for him to act swiftly and surely with the knowledge that we always, always, always cut the red wire? *No, wait! I mean, blue wire!* (Kaboom!)

Ultimately the problem is that we have a hard time paying for knowledge and acquired skills. It's hard for us to account for the years spent learning and honing those skills and factor them into what we're willing to pay. All we see is that we're paying a lot for a task that didn't seem too difficult.

The growing trend of restaurants and artists offering a "pay what you want" model also illustrates how fairness and effort influence our valuations. One restaurant that asked people to pay what they wanted for a meal found that people paid less than the restaurant would have charged normally. That might not sound good for the restaurant owner, but more people came to dine at the restaurant and almost no one paid nothing or very little. In total, the restaurant made more money.[5] This relatively high willingness to pay was likely because people could see the effort—servers taking orders, chefs in the kitchen, food being prepared, changing of linens, and uncorking of wine—and felt the need to reciprocate. To eat at a restaurant and simply walk out without paying seems not only dishonest, but unfair. This scenario also shows that fairness works both ways.

Imagine if, instead of at a restaurant, the pay-what-you-want model had been at a half-empty movie theater. When the movie ended, the theater workers asked their patrons to put in a collection box as much as they were willing. In this case, customers would have felt like it cost the theater nothing extra to have them sit in an otherwise empty seat. They wouldn't have required any brighter projection or better acting.

The theater wouldn't seem to have incurred any extra costs or put forth any extra effort. Thus the theater wouldn't have expended any extra effort and didn't deserve any extra money. The moviegoers would have likely paid very little, if anything at all.

Similarly, people don't feel bad about downloading illegal music and movies for free because they reason that all the effort of producing them took place in the past, and a download does not create any additional effort or cost on the producer's part. (This is why so many antipiracy efforts have focused on trying to highlight the harm caused to writers and performers, in order to personalize the losses.)

The theater/restaurant distinction highlights the problem of fixed versus marginal costs in regard to fairness and effort. Fixed costs, like the seats and lighting in a theater, don't activate our reciprocity as much as marginal costs, like the fresh fish and vegetables the chef grills for us or the shattered glasses from a clumsy busboy's tray that cause people to applaud obnoxiously. (Stop that, people. It's rude.)

The theater/restaurant difference also shows that while we punish prices that we deem unfair because we don't see the effort, we also reward businesses that seem fair by virtue of their conspicuous effort. Isn't this just another example of how we value things in ways that have little to do with actual value? Yes, and that brings us to the issue of *TRANSPARENCY.*

Transparent Effort

James's company didn't blink at paying Gina's consulting firm $725,000, because they appeared to have done such a thorough job, not just of assessing and addressing the company's needs, but of creating a presentation to demonstrate just how hard they worked to do it.

Maybe if the locksmith hadn't offered James so much sass, but an explanation of all the delicate and vital things he had to master and set in order to open the door, the two wouldn't nearly have come to blows. Perhaps if Coca-Cola had explained that it costs much more to keep drinks cool when it's hot, or that someone must drive extra to restock the machines more often on sunny and warm days, people might not have made such an uproar. Maybe then James and Coke consumers would be willing to pay more and be less upset. Because the effort would have been more evident. Any of these would have created a higher level of transparency.

Imagine we have two traditional windup watches, but one has a clear casing so we see the gears grinding in the intricate watchworks. Would we pay more for that watch just because we see how hard it's working? Maybe not (we never did try this experiment), but what is clear is that this is how we unwittingly conduct many financial transactions.

We are willing to pay more when we see the costs of production, people running around, the effort involved. We implicitly assume that something labor-intensive is worth more than something that isn't. It is not objective effort so much as the *appearance* of effort that drives the psychology of what we are willing to pay.

Is this rational? No. Does this warp our perception of value? Yup. Does it happen all the time? You betcha.

The consulting firm that visited James's widget factory did everything short of reenacting their entire project to show the company just how much work they had done. On the other hand, think about similarly expensive law firms that charge an hourly rate. Lawyers are reviled, perhaps, in part, because we don't see the effort that has gone into their work. We just get a bill with hours. Usually more hours than fit in a day, but still, just hours. We see no effort, no tangible sweat, and nothing like the activity the clever consulting firm showed.

Transparency—revealing the work that goes into a product or service—allows a company to show us that they're working hard, earning our money. We don't value things much unless we know there's a lot of effort involved. This is why the Internet is such a challenging medium over which to buy and sell services. Online, we don't see any of the effort involved, so we don't feel like we should have to pay much for apps or Internet services.

Companies big and small have come to learn that transparency shows effort and thus shows—and proves—worth. More and more often, they are working to provide cues to make us value their services more. The travel site Kayak.com is explicitly heavy on transparency. Kayak's website shows us progress as it searches flights, with a moving bar, scrolling items, a growing chart populated with changing options from time to price to airline, making us aware of all the different features being searched. Kayak shows us that a lot of factors are being considered and a lot of calculations are being made. At the end, we can't help but be impressed with all the work being done on our behalf, and we realize that without Kayak, it would have taken us forever, maybe longer, to do all of this ourselves.

Compare this to Google search. We type something and immediately get our answer. What Google does must be simple and easy, right?

Another example is the most innovative change to the pizza industry: the one and only Domino's Pizza Tracker®. Anytime we order Domino's pizza online, a progress bar shows us the changing status of our order—from placing the order to milking the cow for cheese, spreading that on the pizza, putting it in the oven, getting it in the car, weaving through traffic, clogging our arteries, and getting a prescription for Lipitor. Obviously, Domino's skips a few of these steps in its efforts to streamline its Pizza Tracker, but the steps the pizza chain does show attract many people to its website every day to observe the progress of their own pizza.

Some of the most opaque processes are those of the government. One clever project that tried to make government activities more transparent was in Boston. Road repairs in Boston have been going on since the invention of travel. To make the road repairs more transparent, the city government posted online maps of all the potholes that workers were fixing and planning to fix. That showed residents that the city workers were toiling away, even if the road crews hadn't yet shown up in their neighborhood. Boston residents could relate: Now they all understood why it was so hahd to pahk a cah in Havahd Yahd.

Speaking of Bahston, our Havahd friend Mike Norton came up with other creative ways to show the value of transparency, including examples of a dating site that doesn't just show us our compatible matches, but also shows us everyone with whom we're *not* well matched. By showing us thousands of poor matches (let's be honest—they're usually hilariously *horrible* matches), the site's operators also prove how much effort they put into sorting out all the people who signed up for the website—and finding just the right ones.[6] Have we mentioned how we are scared senseless by the modern dating world and how lovely our wives are?

Had Uber, the locksmith, and the umbrella man explained the effort that went into their prices, those explanations may have made the prices seem more fair. Netflix could have explained that there are very high licensing fees for streaming; that the company is lowering the cost for stand-alone users; that Netflix can focus on improving each service; and that it will deliver fresh new programming . . . but it didn't. Restaurants could post signs explaining reasons for every price increase—the cost of gas, raw materials, eggs, labor. They could deflect blame by pointing a finger at taxes or at someone they don't like in the White House. Any of these accounts and explanations would help customers understand and accept these price increases. But businesses don't often do this. Yes, transparency helps us understand value, but, sadly, if we're running a business, we typically don't expect that explaining the effort behind our product or service will change the way customers evaluate it. But it does. . . .

While emphasizing the human desire for transparency helps us see value in the world around us, it also leaves us susceptible to manipulation. The consulting firm demonstrated a lot of effort, but did it really achieve much? The fumbling locksmith worked hard to get the door open, but did he just waste an hour of our time? Are the city workers in Boston really working hard, or just getting dialect coaching?

We can fall victim to transparency, or the lack thereof, more often than we'd like to admit. When we're shown effort, we tend to overvalue a product or service. Transparency, because it reveals effort and thus the appearance of fairness, can alter our perception of value in ways that have little to do with actual value.

Effort Around the House

Our sense of fairness and effort transcends the financial realm. Of course, we can't advise anyone on their personal relationships, but we have found that if we take any couple, put them in separate rooms, and ask the wife and the husband to tell us how much of the total amount of housework they each do, the total always adds up to well over 100 percent. In other words, they each believe they're putting in a great deal of effort, that their partner is doing less, and that, perhaps, that division of labor is not fair.

Why is it that the amount of effort is always more than 100 percent? It's because we are always in the transparent mode. We always see the details of our own effort, but we don't see the details of our partner's effort. We have a transparency asymmetry. If we cleaned the floor, we notice it and we know how much work it took, but if someone else cleaned it, we don't notice the clean floor and we are unaware of the effort that went into making it shine. We know when we take the trash out and all the steps it takes and mess it makes, but we don't see when our partner does it. We know when we put the dishes in the dishwasher using perfect geometric logic and when our spouse just shows no respect for the way plates were obviously intended to fit next to the bowls!

Should we then take the consultant's approach to our relationship, creating a PowerPoint every month to show our partner and kids how many counters we've wiped, dishes we've cleaned, bills we've paid, diapers we've changed, garbage we've taken out? Should we take the lawyers' approach and simply provide a bill detailing hours worked? When we make dinner, should we describe all the steps—from shopping to chopping, cooking to cleaning? Or should we just make a lot of deep sighing sounds—so our spouses will value us more? Well, annoying our spouse with pettiness has its own drawbacks, so we'll let everyone choose the right balance between showing effort and annoying their significant other, but at least take this as some food for thought. Also, remember: Divorce attorneys are expensive. They charge by the hour, and they don't show any of their effort.

FAIR WELL

People always demand what's "fair." In negotiations, sales, marriage, and life. That's not bad. Fairness is a good thing. When in 2015 Martin Shkreli suddenly raised the price of the lifesaving drug Daraprim from $13.50 to $750—that's 5,555 percent—right after he acquired the company that made it, people were outraged. That was seen as blatantly unfair, and while Daraprim remains overpriced and Shkreli remains an [expletive deleted], it has brought long-overdue attention to fairness in drug pricing. So our sense of fairness can be useful, even in the economic world.

But sometimes we overvalue fairness. In less egregious circumstances than Shkreli's, when a price seems unfair, we try to punish the price setter, and we often end up punishing ourselves by passing up an otherwise good value.

Fairness is a function of effort and effort is shown through transparency. Since the level of transparency is a matter of producer strategy, marketing the use of fairness (and especially deceptively promoting our use of fairness) as a proxy for value may not always be done with the best of intentions.

Transparency builds trust and creates value by showing the effort that we connect to fairness. Might unscrupulous people try to take advantage of our desire for transparency and make it *seem* like they worked harder than they *really* did just to add value to their product? Well, in the 150-plus years of hard labor it's taken us to write this book, we have to say . . . no. That would never happen.

10

WE BELIEVE IN THE MAGIC OF LANGUAGE AND RITUALS

Cheryl King is working late. She is spearheading a feasibility study about hiring a team of experts to determine exactly what widgets her company should be making and whether anyone will buy them. No real decisions yet, but she has a deadline and an anxious CEO and no choice but to get it done. She can put up with the occasional late night. What she can't put up with is the awful sushi that comes with the occasional late night.

Every now and then, her team orders this sushi from an allegedly well-reviewed French-Asian bistro downtown called Oooh La La Garden. The trendy restaurant just started delivering. The first time her team ordered from there, Cheryl didn't even look through the menu—in a hurry, she asked her colleagues to choose for her. Her coworker Brian brought her the "Slithery Dragon roll." Cheryl flipped the roll onto a paper towel and absentmindedly started shoving it down her throat while staring at her computer screen. "Yuck," Cheryl thought with her last bite. "Gross. Crunchy and soft at the same time. Oh well."

Meanwhile, in the next room, her coworkers were raving about their meal—whooping, toasting, oohing and aahing. They loved it. Cheryl put on her oversize headphones and tried to focus on the widgets.

Brian soon returned with a bottle of wine. He offered Cheryl a glass, saying he had received the same wine as an anniversary gift, and that it was just amazing. A 2010 Chateau Vin De Yum Pinot Noir—supposed to be excellent. Brian pours some into Cheryl's "One of the World's 500 Best Moms" mug (her kids think they're hilarious). Cheryl takes a sip and mutters, "Uh-huh, thanks. I'll just have a little because I need to get home at some point." Over the next thirty minutes, Cheryl sips from the mug as she wraps up her portion of the project. The wine is okay. Nothing special. Nothing like the wine waiting for her at home.

On her way out of the office, she passes by Brian, tosses him $40 for the food and drink. "We good?"

"Yeah, that covers it. Wasn't it great? You know, it was made with—"

"Yeah, it was fine. See you Monday."

That weekend, Cheryl and her husband, Rick, are strolling down Laurel Street to Le Café Grand Dragon Peu Peu Peu, the new fusion hot spot, whose name sounds like a French machine gun. *Peu peu peu.* Their friends have already arrived, so they slip into their waiting seats.

"Oh my, look at this menu! It's beautiful."

"I know, right? I've heard everything here is good," their friend Jennifer Watson agrees.

Reading the menu, Cheryl coos, "Oooh, look at this: Locally sourced artisanal aged goat's milk fromage graces hand-crafted grass-fed bovine composite mixture complete with fresh-harvest garden yield, heirloom vine-ripened 'tomate,' curated greens, hand-selected onions chosen from a crop of thousands, and special reserve spice blend, imported from global sources and parsed for variance by expertologists, served in the style of a mysterious dark tavern."

"That sounds interesting," Rick says.

"Sounds like an expensive cheeseburger to me," Bill Watson grunts.

The couples chat for a few minutes until their waiter arrives and delivers his modern Shakespearean monologue on the specials of the day. Pointing to the menu, Bill Watson asks him to explain the *spécialité du maison.*

"It means 'house special,' sir."

"Yes, I know, but what is it?"

"Well . . ." The waiter clears his throat. "The chef is very well known both here and in his native France for creating a unique culinary experience for each season."

"Okay, so what is it?"

"Well, this season, it is a filet painstakingly prepared in such a way as to bring forth the flavors of the feed, raised on prairie air, water, and sun and impeccably cared for and curated from birth to the plate."

"Hmm. I'll stick with the fromage thing."

Soon the sommelier comes over and offers Rick the wine list. Heavy book, fine writing. Rick is no wine expert, so he asks for a recommendation.

"Well, the 2010 Chateau Vin De Yum Pinot Noir is the product of an outstanding and rare special harvest. The rains in southern France that summer caused groundwater to be swollen such that the lower portion of most vineyards was inundated with a lush sediment that gave the grapes a fuller, more robust charisma. Pulled from the vine a precisely calibrated 144 hours later than normal and matured using mountain breezes and freshwater, the vintage has received several awards and commendations around the globe. It is intended for an impeccable palate."

Murmurs of general approval. "Sounds great. Let's start with that."

The sommelier returns and pours a splash into Rick's glass. Rick lifts it up, examines it in the light, swirls it in his glass, takes the tiniest of sips, closes his eyes, purses his lips, and swishes it about in his mouth, taking care to wiggle his cheeks. He swallows, pauses, then nods for all the glasses to be filled. Soon they all raise their glasses, Rick makes a toast, everyone chimes in, and the meal is off to the races.

They all share a special appetizer of the day. "This is our famous Slithery Dragon. This roll is hand-crafted with several kinds of chef-select fish like salmon, masago, yellowtail, and tuna belly, all locally raised and harvested, sprinkled with tobiko, scallions, soy-seasoned seaweed, cucumber, avocado, and nuts, freshly washed and wrapped with silvered tongs."

"Mmmmm."

"To die for."

The bill comes. All in all, the wine, the roll, a fancy cheeseburger, and a night of laughter and tall tales comes to $150 per couple. They think it's a bargain.

What Is Going On Here?

These two scenes show us the value-shifting magic of language. Language can shape how we frame our experiences. Language can make us pay extra attention to what we consume and direct our attention to specific parts of the experience. It can help us appreciate our experiences more than we might otherwise. And when we get greater pleasure from something—whether from the physical experience of consuming it or from the language describing it—we value it more and we're willing to pay more for it. The physical thing itself hasn't changed, but our experience of it has and so has our willingness to pay for it. Language is not just describing the world around us; it influences what we pay attention to, what we end up enjoying and what we don't.

Remember the sushi and wine that Cheryl barely noticed in her office? She valued the exact same food and drink much more highly when she became engaged in the language describing them. Similarly, had Cheryl simply had a "cheeseburger" at the restaurant instead of a "locally sourced artisanal fromage bovine composite mixture," she would have enjoyed it much less and would have balked at paying its high price.

Now, of course, eating with friends, far from a computer screen and consultant memos, has additional value in itself. We'd all pay for that. We enjoy the food itself more when it is connected to this type of experience . . . and we are willing to pay more for it. But we can enjoy food more even when the environment is the same, and even the food is exactly the same, if that food is just described differently. Language has the magical power to change how we view food, to get it to command a price that fits the way it is described.

When it comes to creating value, the restaurant environment (luxurious), the social situation (wonderful friends), and the description of the food (all of these postmodern terms) all enhance the experience.

That language is the most powerful, value-raising component of this entire scenario should be clear. Words shouldn't make the seat more comfortable, the spices tastier, the meat more tender, or the company more pleasant. Objectively, it shouldn't matter how an item is described. A burger is a burger, a brownstone is a brownstone, a Toyota is a Toyota. No amount or style of phrasing fundamentally changes what something is. We're either getting a burger, a brownstone, and a Toyota, or we're getting chicken, a condo, and a Ford. We're choosing between things, aren't we?

Well, no. From the early days of research on decision-making it has become clear that we choose from among descriptions of various things, not from among the things themselves. Herein lies the value-shifting magic of language.

Language focuses us on specific attributes of a product or experience. Imagine two adjacent restaurants. One offers a burger that's listed as "80% fat-free beef." The restaurant next door has similar offerings but describes its burger as "20% fat beef." What happens now? The data shows that the two different ways of describing the same burgers cause us to evaluate them very differently. The 80 percent burger focuses on the "fat-free" part, directing us to focus on the burger's healthy, tasty, and desirable aspects. The 20 percent burger focuses our attention solely on its quantity of fat—and therefore we think about its unhealthy aspects. The latter makes us think the burger is disgusting and causes us to look up the rules for veganism. We value the "fat-free" burger much more, and are willing to pay more for it.

The flick of the tongue can be like the flick of a switch, introducing new perspective and context. We've seen that people say they could retire on 80 percent of their current income but that they couldn't retire on 20 percent less than their current income; that we donate to a charity if the amount is described in terms of pennies per day but not when we are asked to donate the same amount described in terms of dollars per year;[1] and that $200 "rebates" send people to the bank, whereas $200 "bonuses" send them to the Bahamas.[2] The 80 percent of income, the charity donation, and the $200 are the same amounts no matter how they're described, but the descriptions change our feelings about a product or service and, as we shall see, change our actual experience consuming them.

The leading practitioners of language manipulation may be winemakers. They have created a language all their own. They use words like "tannin," "complexity," "earthiness," and "acidity" to describe the taste of wine. There are also terms to describe the process of making the wine and how the wine moves, like the "leg" of wine when we swirl it in our glass. It's not clear that most people can either differentiate or understand the distinctions or importance of these items, but many of us act as though we do. We pour wine carefully, we swirl it around, we look at it in the light, and we taste it gently. Of course, we're willing to pay much more for a well-described wine.

On the one hand, paying more for the description of the wine and the process is irrational: The language doesn't change the product. On the other hand, however, we actually are getting more out of the well-described wine. That is, language changes how we experience and consume the wine, influencing us in a deep way but without changing the physical drink in the bottle. The language tells us a story. Then, listening to the description from open to pour, from tipped glass to inhaling the "nose," from swallow to aftertaste, we join the wine's story. That enhances and transforms how much we value the wine and our drinking experience.

So, while language doesn't change the product, it does change the way we interact with it and the way we experience it. Language can also persuade us, for example, to slow down and pay close attention to what we're doing. Imagine we have the best glass of wine in the world, but like Cheryl, we have it while sitting in front of our work computer, paying no attention. How much would we enjoy it? On the other hand, imagine we have an inferior wine, but we think about it, consider its history, taste it, examine it, and cherish it. Despite its objective inferiority, we would get considerable value out of it, potentially more than the objectively better wine.

The coffee industry, like the wine industry before it, has begun to employ creative writers to enhance the language surrounding its product and increase its value. Or so it seems. We hear about "single-bean coffee," "fair-trade coffee," "coffee that has been naturally pressed in the intestines of a cat," the "civet coffee" (you don't want to know), and "coffee that's been sun-kissed by the tears of indigenous people holding the leaves of a thousand generations." That last one isn't true, but it's believable because there's a long, melodramatic story attached to every drop in our Veni, Vidi, Venti cups. And with every detail of a story that we lap up, there's an increase in price we're willing to pay.

Chocolate is following in these footsteps, with so-called single-bean chocolate (we have no idea why solitary beans make better foodstuff, but consumers seem to respond) and other increasingly expensive products. There's a company in the United Kingdom that's catering to "chocolate aficionados." It offers subscription services and all manner of immersive chocolate experiences. For a price, of course. (Who doesn't consider themselves a chocolate aficionado?)

How far will this language trend go? Is there a future for creating and marketing "single-cow milk"? Could we get restaurant menu writers to talk about the personality of Betsy from Minnesota, the cow who provided her third offering of the fifth day of the second week of summer for the latte we've ordered? Would it make customers spend more to know that Betsy's mother once contributed to an ice cream cone consumed by our forty-second president or that her journey to Minnesota was aboard the nation's first hybrid tractor-trailer? That her hobbies include grazing, sunning, and not being tipped? Would customers like to see a picture of Betsy as their waiter describes the "fluency" and "lactose-related viscosity" and "bovine texture" of what they're drinking? Since Betsy lives on a circular farm, we suggest everyone swirl their glasses before dunking their cookies into a tall, frosted, handmade glass of her precious milk. That'll be $13.

As we've seen, language changes how we value goods, services, and experiences of all kinds. After centuries of debate, it seems we've finally disproven Juliet Capulet's theory: A rose by any other name would *not* smell as sweet.

Enhancing Consumption

Enjoyment of something comes from both the sensation of the thing—the taste of the food, the speed of a car, the sound of a song—and what is happening in our brain to co-create the total experience of it. We can call this the full consumption experience.

Language enhances or reduces the quality of the consumption experience—and that's the primary reason it so powerfully influences the way we value something, be it chocolate, wine, or a purebred hamburger. One important type of language that does so is called *CONSUMPTION VOCABULARY.* Consumption vocabulary shows up when we use specific terms to describe an experience, like the "bouquet" of a wine or the "sashing" on a quilt. Consumption vocabulary gets people to think, focus, and pay attention, to slow down and appreciate an experience in a different way and then experience the world in a different way.

A one-minute description of a chef's specialty dish not only focuses us on that dish for an entire minute; it also provides context and depth to the dish itself. It concentrates us on the flavors and texture and taste, giving us a nuanced, complex way to think about the dish. We might imagine ourselves looking, crunching, smelling, or cutting. Our mind and body prepare for the experience. When language supports an experience, or anticipation of an experience, it changes and enhances that experience and how we value it.

As they listened to the waiter describe the specials and the wine, Cheryl and Rick became more and more invested in those items; they become more aware of the special qualities that they offered, and the joy and value they were about to experience.

Although hardly the healthiest of examples, McDonald's commercials used to list all the ingredients of its signature item in a song: "Two all-beef patties, special sauce, lettuce, cheese, pickles, onions on a sesame-seed bun!" For thirty seconds, we think about every item we're anticipating eating. The commercial—like its longer cousin, the infomercial—breaks down the experience so we might consider our one bite to include seven different tastes. What sounds better, that mélange of flavors, or just "a burger"?

Copywriters use consumption vocabulary to highlight the parts of our experience they want us to embrace and those they want us to ignore. Don't worry about the cost of these sneakers and how hard it is to become an elite athlete, "Just Do It" (Nike). Forget about the risks of cutting yourself due to the social pressure to appear clean and orderly; using our razors will make you "The Best a Man Can Get" (Gillette). Sure, you're broke, but "There are some things in life money can't buy. For everything else, there's (MasterCard)." Less subtle consumption copy includes "Have a Coke and a Smile," "Finger Lickin' Good" (KFC), "Tastes Great, Less Filling" (Miller Lite), "I'm Loving It" (McDonald's), and the direct and instructive "Melts in your mouth, not in your hands" (M&Ms).

Jeff noticed the odd juxtaposition of consumption vocabulary in a Café Europa in Times Square, New York. Stenciled signs plant the words "relax," "smiles," "ease," "laugh," "enjoy," "aroma," and "savor" in customers' minds, describing the experience the café wants them to have, so they will value their visit more highly. It must work, since people pay $3.50 for a small coffee. Perhaps more useful signs at that location would have said "Ignore honking taxis," "Try not to inhale through your nose," and "Don't buy theater tickets from a man with no pants."

When consumption vocabulary describes not only what we are about to consume but also the process of production, we appreciate the item even more (remember the impact of effort and fairness), further increasing its value to us. We also become more invested in the product by virtue of our engagement with the language. Remember the endowment effect, where just holding an object can increase its value to us through virtual ownership? So, too, taking the time to have a better understanding and appreciation for the construction of something—an Ikea desk or a fine meal—might increase its value to us.

FUNNY YUMMIES

The tendency of restaurants to overdo it with descriptive language has not gone unnoticed by those in the professional mockery game. Two of our favorites include the menu for the fictitious Fuds (www.fudsmenu.com/menu.html) and the Brooklyn Bar Menu Generator (www.brooklynbarmenus.com), which picks random words to complete the menu at a trendy new hot spot.

As a New Yorker, Jeff can attest that it's believable nonsense that sounds a lot like the actual menus of many trendy restaurants!

crafted lime platter with salt & butter skewers	**14**
miniature bluefish with cider ham	**16**
lamb & french kraut frittata	**14**
winter fig with clam	**14**

rice spread	11
expanded artichoke	18
frightened booze	12
sea-salt rye	10
rubbed marrow, sardine & shell bean tartare	14
water pie with ramp toss	14

Unfortunately, those aren't real options, but don't you want to try some frightened booze? Maybe with a side of rice spread and an order of ramp toss?

Words Seem Fair

One other path by which description creates a powerful influence on how we value things is in conveying effort and fairness. As we just saw, such terms of effort are extremely important. Terms like "artisanal," "handcrafted," "fair trade," and "organic" are used not only to signify creativity, uniqueness, political views, and health, but also to signal extra effort. Effort terms tell us that a lot of labor and resources went into a product and implicitly suggest that the product's value is higher than it would be otherwise. And these words add value.

Would we expect to pay more for cheese produced in small batches using time-honored tools and methods or a similar cheese produced en masse, mostly by machines? Obviously, the small-batch cheese takes much more effort to produce. Therefore, it has to cost more, and we'll probably be willing to pay extra for it. But we might not even be able to notice the difference between the cheeses if the language didn't call it to our attention.

The language of effort is everywhere. Too everywhere. Cheeses, wines, scarves, condos. It's all artisan, artisanal, artsy. There are "artisan lofts" and "artisan dental flosses" (really). Jeff once tried to comfort himself during air turbulence by flipping through the in-flight magazine, but when he came across a story about artisanal moonshine, he felt worse.

"Artisanal" means "made by a craftsman" as opposed to a giant factory. Moonshine is, by definition, distilled whiskey made by hand. The "artisanal" conveys no additional meaning (or value). It just redundantly restates the same thing.

As annoying as ubiquitous words like "artisan" might be, what do they do? They imply that a skilled person made a product by hand, and by definition, anything made by hand takes extra effort. Thus, extra money shall be paid. Think about all the terms hinting at the complexity of the process—the effort heuristics—the waiter used to describe the exact same items Cheryl had cheaply consumed at her desk, description-free.

Sharing Is Fairing

What about the phrase "the sharing economy"? Companies like Uber, Airbnb, and TaskRabbit belong to "the sharing economy," a phrase that frames these services in a positive way. Who doesn't like to share and who doesn't appreciate those who do? Who above the age of preschool doesn't think of sharing as a wonderful human quality? No one, that's who.

The phrase "the sharing economy" conjures an image of the good side of humanity, and that causes most of us to value a service more. Certainly the language doesn't draw attention to the negatives of the sharing economy. "Sharing" makes it all seem selfless, like we're letting our little sister play with our Legos or donating a kidney to an orphan. But that's not always the case. In fact, critics claim that the rise of the sharing economy is the by-product of a labor market providing no full-time jobs, few benefits, and little security, that it rolls back worker protections and takes advantage of the "free-agent nation," which itself is another term designed to help us feel better about underemployment. But we do all enjoy getting rides more easily, don't we?

Some companies have been accused of greenwashing, or making minor, cosmetic changes to their products so they can call themselves environmentally friendly. Others have been accused of pinkwashing—paying to be certified as pro–women's health by organizations like the breast cancer advocacy organization Susan G. Komen—because they know we will pay more for products associated with the extra effort to do good for the world. Good marketers are incredibly adept at using language to convey a sense of wonderfulness, yet there really aren't any strict rules regulating who can call themselves "green" or "fair trade" or "good for babies and trees and dolphins." Anyone can create an organization, hire a graphic designer to make a logo, and slap that thing on any product out there. And there you have it: "A Healthy Smart Choice Selection," "Environmentally Friendly," or "Certified by the Council for Good Things That Make You Happy."*

The point is that language offers a window into the effort we so crave to see, which signifies fairness and quality. In turn, perceptions of fairness and quality become a proxy for value. That is the long and windy route we travel from language to value, and we can be tripped up at any step along the way.

Double-Talk

Language can not only create a perception of effort and a sense of value; it can also get us to attribute expertise to the people using these terms. Consider the professions of health care, finance, and law. We laypeople have no idea what some of their phrases mean—medial collateral ligament, collateral debt obligations, debtors' prison—and we often can't even read their handwriting. Obscure and impenetrable language conveys a sense of expertise. It reminds us that they have greater knowledge than we do, that they must have worked hard and long to gain all that knowledge and skill, and now they get to show it to us by using their

*By the way, the book you're reading was certified "A-Plus-Number-One" by the Council of Good Things That Make Your Life Better. Congratulations on a smart and healthy choice.

overly complicated language. Therefore, they certainly must be extra valuable.

This use of language creates what author John Lanchester calls "priesthoods"—using elaborate ritual and language that is designed to bamboozle, mystify, and intimidate, leaving us with a feeling that we are not sure what's being talked about but that as long as we use the service of these qualified people we will be in expert hands.[3]

Again, the sommelier's description of the wine was enticing in its complexity and poetry, but it was also confusing for those who know not of rains and harvests and tannins. It sounded special because it sounded like something only experts understood. Lucky for us, we get to benefit from their hard-won and obscure expertise.

In this case, it's the *lack* of transparency that adds value. Obscurity in winemaking, or in any other process that isn't the province of the layperson, creates a sense of underlying complexity that may not be warranted. But that sense of complexity nonetheless influences how we value the experience itself.

"Up" Is "Down"

We might think that an evocative description can change the value of an experience only incrementally—say, influencing Cheryl to spend $150 on dinner instead of $40. But, in fact, rich, specific, sensory descriptions can quite dramatically change the value of an experience—Cheryl was willing to spend $150 on dinner in the restaurant, in contrast to $40 on the meal in her office. Furthermore, they can even influence whether we're going to pay or *be paid* for a good or service.

In Mark Twain's brilliant book *The Adventures of Tom Sawyer*, Tom must whitewash his aunt's fence. When his friends mock him for having to work, he replies, "Do you call this work?" "Does a boy get a chance to whitewash a fence every day?" and "Aunt Polly's awful particular about her fence." Having heard the work of whitewashing the fence described as pleasure, his friends scramble to experience the joys of whitewashing, and before long trade Tom their favorite personal items for the privilege of doing so.

At the end of the chapter, Twain writes, "If Tom had been a great and wise philosopher, like the writer of this book, he would now have comprehended that Work consists of whatever a body is obliged to do, and that Play consists of whatever a body is not obliged to do. . . . There are wealthy gentlemen in England who drive four-horse passenger-coaches twenty or thirty miles on a daily line in the summer because the privilege costs them considerable money; but if they were offered wages for the service, that would turn it into work, and then they would resign."

Language can be transformative. It can turn pain into pleasure or a hobby into work, and it can make those transformations go in either direction. Jeff claims to reflect about Tom's whitewashing adventure every time he submits something to the *HuffPost*—for free. By all accounts, founder Arianna Huffington was one of the greatest fence painters of all time: She successfully offered "exposure" and, in doing so, demonstrated the magical power of language.

Ritual Me This

How do rituals fit into all of this? Did Rick's swirling his wine, pursing his lips, and raising a toast make the wine taste better than it would have otherwise? Actually, yes—and to a larger degree than we might expect.

Descriptive language and consumption vocabulary for any given product or service tend to be remarkably consistent. It doesn't change very often and it builds on itself. We always think of the same terms for each new experience of a product—the nose of a wine, the texture of a cheese, the cut of a steak. In addition to the value-enhancing benefit of the language we've already discussed, this consistency in these terms—how we use them and repeat them and how they inform the way we behave—creates rituals.

Rituals connect a single experience to many other past and future experiences just like it. That connection gives the experience extra meaning by causing it to become part of a tradition that extends back to the past and forward into the future.

Most rituals come from religion. We have religious rituals like putting on a yarmulke in Judaism, counting the beads in Islam, or kissing the cross in Christianity. Yes, all of those rituals are actions with specific processes and descriptions. They all link people to past actions and to their own history. But most important, they are symbols that convey an extra sense—a higher order—of meaning. And that makes whatever is connected to the ritual much more valuable than it would be on its own—be it a prayer or a glass of wine.

Remember, enjoyment comes from the experience we are having from the external product or service and from the experience we are having in our brain. Like language, rituals enhance the experience of consumption, which, by expanding our sense of connection to past experiences and creating a sense of meaning, increases our enjoyment. In the process, rituals increase our valuation of the thing used in that ritual: A piece of sushi, or a glass of wine, can seem "more expensive" by virtue of the actions we take and the movements we make when we consume them.

Kathleen Vohs, Yajin Wang, Francesca Gino, and Mike Norton studied rituals.[4] They found that rituals can increase enjoyment, pleasure, value, and, of course, willingness to pay. The lucky participants in their study were given a chocolate bar and asked to consume it either by eating straightforwardly, or by first breaking it and unwrapping it in a particular way and only then eating it. Those who broke and unwrapped it in the particular way were essentially performing a ritual before consumption. It wasn't a very meaningful ritual, but it was a ritual nevertheless. Similarly, two other groups were given carrots and asked either to eat them regularly or to first perform rituals that included rapping their knuckles, taking deep breaths, and closing their eyes, and only then eating the carrots. It's too bad, for science, that they didn't think to include the ritual of taking a bite, then asking, "What's up, Doc?" That would have been really great. . . . For science, of course, not just for our amusement.

What they found was that the people who engaged in rituals savored the experience of eating much more. This was true for both carrots and for chocolate. Rituals increased the experience and enjoyment, both in anticipation of the actual experience and in the moment. Surely increased enjoyment is worth something, isn't it? Why, yes. When they tested "willingness to pay," they found that those who ate the ritualized chocolate were willing to pay more and thought that what they were eating was "fancier."

Rituals aren't just weird knock patterns and fancy breathing. They can include almost any action and type of experience. Making a toast, shaking hands, saying grace, or breaking an Oreo cookie in half and licking off the icing—these and so many other rituals help us become more present, so that we focus more on the experience, the item, or the consumption at hand.

The rituals we undertake during consumption make the experience special. We own it more; it becomes a greater investment, one that is more entangled in our own lives and experiences. We also get a greater sense of control through rituals. An activity becomes familiar. It becomes our own when we ritualize it. We are in command. That adds value, too.

Rituals make food seem tastier, events seem special-er, and life seem life-ier. They make experiences feel more valuable. Like consumption vocabulary, rituals make us stop and focus on what we're doing. They enhance our enjoyment of consumption because they give us greater involvement in that consumption. But rituals go a step beyond consumption vocabulary because they also involve some activity on our part, and they also involve meaning. In the process they can enhance almost any experience.

We may drink just one glass of wine, but with a ritual, we experience more pleasure in that moment of drinking than we would without it. Two identical bottles of pinot noir, side by side—one poured into a coffee mug and the other into fine crystal, swirled, held to the light, dripped on the tongue, twirled in the mouth—which should Rick value more highly? For which would we pay more? The bottles and the wines inside them are the same. They should be valued identically. But they are not. We value ritualized wine more! Our spending behavior in this regard is certainly not economically rational, but it is understandable, and in some cases even desirable.

Open Wide

For those of us doubting the consumption-enhancing power of ritual and language, try feeding mashed peas to a toddler.* Then try doing the same thing again but this time by telling the wee one that the spoon is an airplane, coming in for a landing. Swirl your arm in the air. Make the buzzing propeller sound! Go, go, go! We'll look ridiculous, but we know even the stingiest toddler would pay more to eat a tiny little airplane than a spoonful of green mush. If we think that we adults have outgrown the influence of a show on what and how much we are willing to eat, go to a Hibachi-style restaurant or a murder-mystery dinner theater or stop and examine whatever is being shoved into your face-hole while binge-watching must-see TV.

We humans want to believe that our food is going to be delicious, our investments will pay off, we can find a great deal, we can become an instant millionaire, and that we're about to eat an airplane. If that's what language and rituals tell us, we'll suspend disbelief—at least to some degree. We will experience what we want to experience.

Rituals and consumption language influence us to value things more than they are objectively worth. Their magic is in the way they transform our experiences, all the way from purchasing day-to-day products to making large decisions about such things as marriage, jobs, and the ways in which we interact with the world around us.

* Norton suggests that parents have been pretending that a spoonful of peas is, say, "a plane coming in for a landing" in order to make it more appealing *for centuries*.

11

WE OVERVALUE EXPECTATIONS

Vinny del Rey Ray likes the good life. Fast cars, hot deals, fun times. He considers himself a connoisseur of all things fantastic. He's on top of every trend, ahead of every curve, pushing every envelope. If something is considered "the best," he has to have it—and then brag about it. In fact, if something doesn't have an excellent reputation, he won't touch it. He isn't superrich, but he has money, and he can afford to make sure he doesn't waste his life with inferior products and experiences.

He wears Armani suits. The best. They feel good. They look good. They project an air of success that has served him well in his work as a commercial real estate dealmaker.

Today he's driving to sign a real estate deal in his new Model S Tesla—the best car in the world. No emissions. High speeds. Envious looks. Vinny leases a new luxury car every year or two. He'd read all about the Model S before he got behind the wheel, but it was the test drive that sold him. He could feel the power, the handling, and the control that he'd read so much about. He could see the stares and hear the whispers he'd dreamed about. This car was made for him.

Vinny believes himself to be the top real estate negotiator in the Valley. Which Valley? All of them. But today he's going to strike a deal with Richard Von Strong, a man whose reputation for success—and viciousness—precedes him like a shock wave. Normally cool, calm, and collected, Vinny has had a terrible headache all day. He spins his wheels into the parking lot of the first convenience store he passes.

Inside, he searches for some Extra Strength Tylenol. They don't have any. "Here, try Happy Farms Acetaminophen," the clerk offers. "Same thing as Tylenol, much cheaper."

"What? Are you kidding me? Don't give me that cheap knockoff junk. It'll never work. Tylenol does the trick. Thanks anyway, pal."

Back in the Model S, Vinny backtracks a couple of miles, gets his Extra Strength Tylenol, and washes it down with a splash of $3 vitamin water.

Vinny pulls up to the luxury hotel where Von Strong holds all his meetings. Von Strong is notorious for renting out a penthouse suite to intimidate his adversaries. Vinny's head throbs. Rubbing it, he passes up the open parking spots and gives the valet his keys, taking great pains to tell the teenager at the desk that the Model S was the top-reviewed car in its class, performs like the rocket ship of his dreams, and also saves the planet.

In the elevator, Vinny gets a text from his assistant. Apparently, Von Strong had to rush off to a family emergency and his business partner, Gloria Marsh, will take his place. Vinny takes a deep breath, relaxes his shoulders, rubs his silky suit, and feels his headache subsiding.

Vinny's at ease in the negotiation, figuring Gloria can't be as tough as Von Strong. He listens to her first offer enthusiastically, since it is clear that she's surely not the type to play tough. He counters with a figure higher than he was prepared to go with Von Strong. He's not worried; she's not going to get the best of Vinny del Rey Ray. Not today day. In the end, he gets his deal. The terms are less favorable than he'd hoped to get from Von Strong, but he feels good about it.

He leaves, texts his assistant to get the best bottle of wine she can find, and hops in his Model S to go celebrate.

What Is Going On Here?

Vinny's is the story of how expectations distort our value judgments. Vinny expected his car to drive, look, and be perceived as better than any others, so he paid more for it than one with lesser expectations. He expected the Tylenol to relieve his headache better than a no-name brand of the same chemical, so he paid more for it. He expected a man to be a tougher negotiator than a woman, and he paid for that, too.

If we've ever read about the stock market, we've come across "expectations." Stock prices often reflect how a company performs relative to analysts' expectations. A company like Apple may make 70 kajillion dollars one quarter, but if analysts expected it to make 80 kajillion, then it "fell short of expectations" and its stock price will likely fall. So, relative to expectations, Apple will have performed poorly.

But there's a trap here that we tend to overlook. It was the analysts' expectations that raised the stock price too high in the first place. Analysts expected Apple to do very well—80 kajillion well—so they increased their perception of the company's value. This is what our brains do with experiences, too.

Much like a company stock, our own valuations are affected by the expectations of our most trusted analyst: ourselves. If we expect something to be really, really fantastic, we will value it more highly than if we expect it to stink. We'll expect the same wine drunk from fine crystal to taste better than from a cracked mug, and we'll pay for it, too. This is true, even if the fundamental underlying i-gadget, widget, or wine is exactly the same.

The brain plays a big role in the way we experience things. Duh.

The future is an uncertain place. We don't know what's going to happen. Even when we do know the general plan—tomorrow we'll wake up at six thirty, shower, grab coffee, go to work, come home, kiss our loved ones, go to sleep—we don't know all the details, all the unforeseen twists and turns. The high school friend we'll see on the train, the office birthday cake we'll drop on our pants, or the unexpected sexual tension we'll share with Mavis in the copy center. Oh, Mavis, you and your collating . . .

Luckily, our brain is working hard to fill in some of the gaps for us. We draw on our knowledge and imagination to anticipate the details of a future experience. This is what expectations do. Expectations add color to the black-and-white images we hold of our future selves.

Our imaginations are incredibly powerful. Elizabeth Dunn and Mike Norton ask their readers to imagine riding a unicorn on the rings of Saturn (really), then they point out that "the ability to conjure up an image of this awesome and impossible activity contributes to the magic of being human, and demonstrates our ability to go almost anywhere in our minds."[1]

Picture our imagination of the future as a surface, with cracks, crevices, and gaps. Those gaps can be filled in with the gooey fluid of expectations. In other words, our mind employs expectations to complete our vision of the future. Our minds are awesome. It's a shame so many of us try to stab them with *The Real Housewives of Midsized City, USA*.

Great Expectations

Expectations alter the value of our experiences during two different time periods: before we experience a purchase, or what we might call the anticipation period, and during the experience itself. These two types of expectations act in fundamentally different but important ways. Expectations provide us pleasure (or pain) while we anticipate an experience and then they also change the experience itself.

First, while anticipating a vacation, we're planning it, imagining the good times, the fruity drinks, and the sandy beach. We get extra pleasure from our anticipation.

The second effect of expectations, however, is much more powerful. During the experience, expectations actually have the ability to change how we experience the world around us. A week of vacation can become more enjoyable and more valuable because of heightened expectations. We pay more attention and we savor moments more fully as a result of expectations. It's not just our mind that changes because of expectations; our body changes as well. Yes, when we spend time anticipating something, our physiology changes, too. The classic example is Pavlov's dog, whose mental anticipation of food caused him to salivate.

The moment we begin expecting something, our minds and bodies begin preparing for that reality. That preparation can, and typically does, affect the reality of the experience. Woof.

WAIT, WHAT? EXPECTATIONS MATTER?

Unlike the other psychological effects of money we explored so far, expectations—like language and rituals—can change the real value of our experience, not just our *perception* of that value. We'll explore this important distinction more in part 3 of this book, when we suggest how we can actually use some of our human quirks to our advantage.

Anticipation . . . Is Making It Great

In the anticipation period, expectations add value to or subtract value from every purchase we make. If we expect something to be a positive experience, we prepare for that, perhaps smiling, releasing endorphins, or simply seeing the world in a more positive light. The same is true with negative expectations. If we expect something to be bad, our bodies get ready for that negative experience, perhaps by tensing up, growling, getting stressed, staring at our shoes, and girding ourselves to face the miserable world around us.

If we gain pleasure from anticipating a fun vacation, that enhances our experience of the vacation when we get there. If we spend four weeks daydreaming about lying on the beach and drinking cocktails, there's a value in that. If we add the pleasure of expectations to the actual experience—four weeks of dreaming plus a week of actual vacation—we see how expectations increase the overall value to us, above and beyond just the moment of actual vacation. Put another way, purchasing one week of vacation brings us five weeks of pleasure. (Some people say that they buy lottery tickets knowing full well that they won't win, because it gives them a few days of pleasure imagining what they will do with their winnings.)

Similarly, low expectations can decrease the pleasure of an experience. If we have that root canal coming in a week, it can ruin every day leading up to it, with all the horrible images and nightmares we'll have. Then we'll have the root canal. And it will hurt. We'll have root canal pain plus the root canal dread, which is not fun, even though it does sound like a great name for a punk rock band. (Tonight Only! Root Canal Dread . . . *You Know the Drill!*)

Remember how we discussed that rich descriptions and rituals enhance the "consumption experience"? Expectations operate in a similar way. Enhanced expectations change how we value experiences themselves. Expectations act as value cues that aren't tied directly to the thing we are buying. They are not changing the purchased item; they are our brain's perception of that item, which changes how we experience it. . . .

The Expectation–Experience Connection

It's not just our perception of something that gets changed by expectations, but the actual performance and experience of the thing itself. Expectations have a real impact, not just on how we prepare for an experience, but in what that experience subjectively and objectively feels like.

Expectations have been shown to improve performance, enhance the consumption experience, and change our perceptions, thereby affecting our ability to assess value and willingness to pay. Like language and rituals, expectations help us focus on the positive—or negative—

aspects of that experience, thus giving those elements lots of weight. From wherever they may come, expectations have the power to change our reality.

Vinny expected his Tylenol and Tesla to work well, so, in his experience of them, they did. People who expect a cartoon to be funny laugh more; those who expect a politician to perform well in a debate believe he or she did[2]; and those who expect a beer to taste bad end up not liking it as much as they would without such expectations.[3]

In Rudolf Erich Raspe's classic, *The Surprising Adventures of Baron Munchausen*, the tale's hero is stuck in a swamp. He gets himself and his horse out of the muck by simply pulling up on his own hair. While this is, of course, physically impossible, Munchausen believed it would work—he *expected* it to work—and so it did. Unfortunately, we nonfiction characters aren't able to use expectations to change our bodies that much, but still, they do make a difference.

There is a good deal of research on how expectations change performance of our mental activities. Some of the most surprising—and disturbing—findings include the following:

A. When you remind women that they are women, they expect to perform worse on mathematics tasks and they actually do perform worse on those tasks.
B. When you remind women who are also Asian that they are women, they expect to perform worse on math tasks and do. But when you remind the same women that they are also Asian, they then expect to do better on math tasks and they actually do perform better.[4]
C. When schoolteachers expect some kids in the class to do better and some to do worse, each group of kids performs up to, or down to, those expectations. This was because of the way the teacher's behavior and the children's expectations for their own performance was shaped by the teacher's initial expectations.[5]

While these studies do have wider implications on the impact of stereotypes and biases, for our purposes, they simply emphasize the ability of expectations to alter our mental outlook and abilities.

It's worth noting that there is a growing cross-cultural embrace of the power of expectations to impact performance beyond just our *mental* abilities. From Oprah Winfrey's plea to "put it out into the world" to the spread of "vision boards" and the use of—and die-hard faith in—visualization by elite athletes, people believe in the transformative power of creating expectations. While we aren't going to comment on the scientific efficacy of these particular practices, we—authors of what will be a worldwide bestseller, a major motion picture, and a key to advancing life and peace on earth—believe in it some, too.

So, expectations matter, but where do they come from?

BRAND-NEW YOU

Branding creates expectations because branding increases the perception of value. Branding Works!!©®™ It certainly influences subjective performance, as studies going back to ancient times—that is, the 1960s—confirm. The same meat[6] and beer taste better when there's a brand name attached.[7] And, to get all neuroscience for a moment, people "reported greater pleasure as they consumed Coke-branded cola, corresponding to higher activation levels in the dorsolateral prefrontal cortex, an area of the brain associated with emotions and cultural memories."[8] In other words, branding doesn't just make people *say* they enjoyed things more; it actually makes these things more enjoyable inside their brains.

In a recent branding study, people with too much time on their hands—also known as "volunteers"—were asked to try out products, some of which carried fancy brand names and some of which didn't. Participants wound up truly believing that brand-name sunglasses blocked out more glare than lesser-known ones and that brand-name earmuffs canceled out more sound. In these experiments, all the products were the same; they were just branded differently. The label made a real impact on the perceived usefulness of each product.[9]

We might expect brand names just to improve expectations—that a product *would* block more light and silence more noise. But in fact, the expectations created by the brand names actually improved objective performance: When we examine real performance, we see that the brand-name product *did* block more light and silence more noise. Participants preached themselves into becoming true believers, converts to the church of the holy brand. They expected brand-name items to perform better, to be of greater value, and it was their very expectation of such increased value that made it so. It was a self-fulfilling sunglass and earmuff prophecy.

We also like to stick to brands we've come to trust. Maybe we've always bought a certain type of car—say, a Honda. We believe that brand has greater value than others, that it must be better, that our judgment must be right. Dick Wittink and Rahul Guha found that, indeed, people who purchase a new car from the same automaker as they had before pay more than those who are buying that brand of car for the first time.[10] It's a self-herding* and a name-brand premium combined.

Reputation—related to, and often confused with, brand—also shapes expectations. We see the effect of reputation everywhere.

It wasn't just the names of Tesla, Tylenol, and Armani that made Vinny believe that his chosen items were faster, more prestigious, finer products. It was also the reputations of these particular products.

Dan and his colleagues Baba Shiv and Ziv Carmon conducted an experiment in which they presented participants with Sobe energy drinks, either on their own or along with literature that claimed it improved mental function and puzzle solving. The participants who received the literature also received many (fictional) scientific papers to support this claim. What the results showed is that the participants who got all the (fictional) studies performed better on subsequent tests than those

* Flip back to chapter 7, "We Trust Ourselves," for a refresher.

Sobe drinkers who didn't get the scientific stamp of approval. That is, the reputation of Sobe-as-problem-solver gave the study volunteers an expectation that drinking it would boost their mental performance, and that expectation led to actual improved performance.[11]

In July 1911, the *Mona Lisa* was just another painting. In August 1911, it was stolen from the Louvre. While the authorities tracked it down, there were suddenly huge lines of visitors waiting to view the empty space where the painting had hung. More people went to see the absence of the painting than had gone to see the painting itself prior to the theft.

The theft had become a signal of the *Mona Lisa*'s worth. Surely, no one would steal a worthless painting. The crime brought long-term value to the *Mona Lisa* and the Louvre. These days the painting might be the most well-known piece of art in the entire museum. The painting's value is immeasurable. Its reputation—created through theft—now precedes it worldwide.

Jeff went to Princeton, a "prestigious" and "highly regarded" university that provided him four years of "beer" and "pizza." He expected an excellent education, he probably got it, and he certainly paid for it. He also has reaped the benefit of the school's reputation—regardless of how much book learning he may have done—from job interviews to professional networking to lines at tailgate parties. The reputation of a wide range of schools often shapes the expectations of everyone from parents to admissions officers, job recruiters to blind dates. This isn't to say their reputations aren't deserved, but their brand and reputation certainly affect people's opinions, and expectations, of their graduates.

The Past Is Prologue

Our past experiences also shape our expectations about future experiences. A good experience with a product—a car, a computer, a coffee, a vacation destination—will make us overvalue that product by projecting our past experience on to our potential future consumption.

Hollywood pumps out sequels and remakes galore. (Studies would probably show that 145 percent of all new Hollywood projects are just old Hollywood projects with new names.) Why? Because we liked the original film and rewarded the studio at the box office. Because our collective previous experience was good, everyone's expectations for the follow-up (especially the studio) should be high. At least high enough for me to fork over $15 to see them ruin my childhood.

One of the problems with the expectations that result from past experiences is that if they are too divergent from the experience itself, it can set us up for disappointment. When the contrast between expectations and reality is too large, the force of expectations cannot overcome this gap and high expectations backfire. JCPenney customers expected sale prices, so when they didn't see sales, they were outraged, even though the actual prices were functionally the same as before.

Imagine a teenager who gets a $25 gift card as a birthday present from an aunt who, for many years past, had sent $100 gift cards. What would his or her reaction be? "She normally sends $100. This sucks. I totally lost $75. Something-something Instagram Snapchat social media OMG!" Rather than seeing it as $25 gained, she views it in terms of her expectations of $100 based upon past patterns and perceives the gift as a loss.

Once again, past performance is simply no guarantee of future success. But go tell this to our expectations. Just because something went well in the past, that doesn't mean it will in the future. A steak might be overcooked, a hurricane might hit our tropical vacation spot, a scary moment in a horror flick might seem trivial without the element of surprise. We only get one chance to make a first impression; this is true of people and purchases. But our expectations don't work this way. They are preloaded with our past experiences, eager to be applied over and over to the same experiences and to new ones.

Presentation and setting also create expectations that help make perception become reality.

Pouring wine into glasses of different shapes, styles, and materials—a shot glass, a fancy crystal flute, a mug—can change the perception of value and at the same time change the price of the wine. Remember when

Cheryl drank the fine wine in a coffee mug at her desk and then later in a nice restaurant setting with friends? The liquid—the same product—was worth much more to her when she drank it out of a delicate crystal glass.

Marco Bertini, Elie Ofek, and Dan ran an experiment in which they gave coffee to students. They placed condiments nearby in either fancy dishes or in Styrofoam cups. Those who got their cream and sugar from the fancier setup said they liked the coffee more and would pay more for it, even though, unbeknownst to them, it was the same coffee as the one served near the Styrofoam cups.[12]

Similarly, a famous virtuoso playing violin in the subway sounds like a pauper to those rushing by, while an amateur raking the strings in an opulent state theater sounds, well, not "good," but not as bad as he or she would on the street.

Timing Is Everything

The power of expectations is more potent when we pay for something before we consume or experience it.

As an example, let's revisit the pain of paying. When we pay before we consume something, it reduces the pain we feel at the time of consumption. If we pay $100 for something that we won't consume for, say, three months, we get the $100 thing, plus the three months of anticipation and daydreaming and excitement. So we get more than we pay for, and when we finally get to consume the thing, we might even feel like we're getting a bargain.

Paying after consumption also reduces pain at the time of consumption to some degree, but we get less value and less joy from anticipating the consumption experience itself. When reflecting upon the past, we must use our memory, which, with those stubborn facts and details, has less creative freedom than our imagination does when we use it to dream about the future, with all of its blank spaces and beckoning possibilities. Darn you, memory!

University of Southern California students got more pleasure from a video game if, before they played, they imagined how awesome it was going to be. Delaying consumption increases what social scientists call the "drool factor." Using chocolate and soda, they found that participants enjoyed consuming these things more if they had to wait a while to do so.[13] While these results reinforce what we instinctively know about the increased pleasure that results from anticipation, it seems that someone needs to figure out why so much of social science involves chocolate-based experiments.

Remember how Jeff and his wife paid for their honeymoon in advance and got several weeks to imagine how much fun it would be? That showed the benefit of expectations of a pleasurable experience. On the other hand, negative expectations can lower our valuations. Dan and his colleagues once gave students beer laced with a dash of vinegar. (There was just a little vinegar in the beer, but enough to change the beer's taste.) Those they told about the vinegar *before* they drank the beer enjoyed it much less than those who learned about the vinegar *afterward*. If we tell people that something might be distasteful, it's likely that they will agree with us not just because the physical experience is different, but also because of the expectations brought on by the warning.[14]

The future holds endless possibilities. When it comes to those possibilities, we tend to be optimistic. Anticipation, imagination, expectations—all these things contribute to increasing the value of whatever we'll get later, whether it's a show, a trip, or a gourmet chocolate delight. However, when we reflect back upon an experience, reality rudely guides our evaluation. We are forced to fill in the blanks with facts. Unless we're a politician, but that's a discussion for a different time.

Knocking and Talking Revisited

Rituals and language also create expectations that impact performance and enjoyment. We have already discussed the ways in which detailed descriptions—say, of items on the menu at fancy restaurants—increase our attention and focus. But we haven't yet unpacked how they also increase our expectations. Any meal worthy of a three-minute monologue must be delicious. That's what we expect, and that's what we'll convince ourselves we're experiencing.

We know that rituals can further enhance our experience. They reduce anxiety and increase confidence, focus, and attention.

In *Predictably Irrational*, Dan described the ritualistic benefit of Airborne, the dietary supplement that claimed to prevent or cure the common cold. The fizzing and foaming that made it feel like it was working. That ritual influenced him to focus and caused him to expect to feel better. Before performing onstage or playing pool, Jeff goes through certain rituals—with chewing gum, Tic Tacs, and ginger ale (don't ask). Are these superstitions or rituals or just silly? We don't know. We do know that he believes they make him do better—maybe because he grew up inspired by the quirky rituals, and undeniable success, of Boston Red Sox oddball third baseman Wade Boggs.*

Expectations? Great!

We've barely scratched the surface of the many origins of our expectations, but the point is to realize how common and powerful they are. Their impact is undeniable: They make us value things in ways unrelated to actual value, and they are everywhere.

It is clear that expectations change how we value things in life, from the mundane (Tylenol and coffee) to the sublime (art, literature, music, food, wine, companionship). If we have high expectations for an experience, regardless of the source of these expectations, we will value it highly and be willing to pay a premium for it. If we expect less, we'll value it less and be willing to pay less. Sometimes this is good. If we

*Boggs—a five-time batting champion—ate chicken before each game, scratched the Hebrew word for "living" into the dirt before each at bat, and had a bunch of other specific rituals, like the timing of batting practice, stretches, and fielding practice. He was awesome. It's a shame he had to go play for the Yankees, or, as it's known in New England, "get run ovah by a cahr."

are going to love our sushi more, maybe we should pay more for good expectations and more enjoyable sushi. But sometimes it's not as clear. If we believe an expensive name-brand product works better than the same product with a generic name—and our expectations make it so—should we pay more for it?

Some of us rely upon our expectations more than others. We admit, Vinny seems like kind of a jerk (apologies to the Vinny-American community for the stereotype). Hopefully the rest of us are not jerks, but we are like Vinny sometimes, when we, in our failure to recognize our behavior, rely upon our expectations to evaluate our choices and determine our spending.

Of course, a powerful source of value-shifting expectations is the very thing we're trying to figure out: money. When things are expensive, we expect more from them, and when they are cheap, we expect less. Then, through a self-created cycle of expectation and value, we get what we (are willing to) pay for.

12

WE LOSE CONTROL

Rob Mansfield will be able to retire shortly after pigs fly.

A highly educated, successful, self-employed businessman, Rob has not been saving for retirement. In his twenties and early thirties, he worked at a large company that offered a retirement plan, including corporate matching, but he chose not to enroll. Making what he considered a meager salary, he felt he needed every penny just to scrape by and have a little fun while he was young enough to enjoy it. Choosing to take a few hundred dollars out of his paycheck seemed like a dumb idea to him. Instead, he chose to live it up for the next five or ten years. Once he gets a substantial raise, he figures, it will be no problem to save a lot every month. Future Rob will take care of Retired Rob.

As a freelance consultant running his own business, Rob now earns a good amount of money. It's not consistent, but he's able to pay the bills for himself and his new wife, as well as occasionally enjoying the finer things in life. Each month he sets aside money for taxes and health insurance, but not for retirement.

At his wedding five years ago, his new parents-in-law entertained

Rob's guests with stories of their early retirement. They'd been frugal savers, and now, just in their early sixties, they were enjoying a simple, but work-free, lifestyle. They traveled to see relatives, played tennis, spent quality time together. Oh, and ate at a lot of buffets.

It seemed deadly boring to Rob, who reveled in the excitement of running his own business and the rewards of dining out, traveling, and buying new toys whenever he landed a new contract. He has an affinity for classic motorcycles. He buys a new one every few years and is constantly upgrading, refurbishing, and polishing the chrome on the ones he has. Sometimes he even rides them.

About two years into their marriage, at the urging of her parents, Rob's wife asked him, for the first time, about his retirement plan. He joked that he'd been investing in the lottery and had recently planted two acorns and bought a hammock.

His wife narrowed her eyes and asked, "Really?"

He responded, "Not really, but don't worry about it."

"Rob!"

"It's fine."

As she stormed out of his entertainment center/man cave, her unprintable response gave him a pretty good money-saving idea: a swear jar. He'd be rich by now.

Since that encounter, each month Rob considers starting a self-funded IRA. But at the end of each month, no matter how much he's earned, he feels like he can't afford it. He has bills to pay. Plus, there are things he wants to do for himself and for his wife—romantic dinners, weekends away, new bike gear, upgraded sound systems—and it is more important for them to feel good and enjoy life while they can than to save. In fact, years have passed, and he's still not saving. And now the work is drying up a bit. Future Rob isn't saving any more today than twenty-five-year old Rob was.

Unfortunately, Rob is in good company in failing to save (or not saving enough) for retirement. In 2014, almost one-third of American adults had not started saving for retirement. And nearly a quarter of those closest to the end of their careers (ages 50–64) had not begun saving for retirement.[1] Put another way, nearly 40 million working-age households in the United States do not have any retirement assets. Even among those who do, account balances are far below conservative estimates for how much these households will need to fund their retirements.[2] Another survey found that 30 percent of Americans are so behind in saving for retirement that they will have to work until they're eighty.[3] Average life expectancy is . . . seventy-eight. That's negative two years to enjoy retirement. We're not just bad at saving, we're bad at math, too.

One interesting survey even found that 46 percent of *financial planners* don't have retirement plans themselves.[4] That's correct: Those whose job it is to help us save are not saving. Good luck, world.

What's Going On Here?

Rob's story—and that of retirement saving in general—highlights our problems with delayed gratification and self-control. We have a hard time resisting temptation, even when we know all too well what is good for us.

Raise your hand if you promised yourself last night that you'd wake up early and work out today. Keep your hand up if raising your hand is all the exercise you've gotten today.

Delayed gratification and self-control are not strictly about the psychology of money, of course, but our ability to delay gratification and to control ourselves influence how we manage (or really, how we *mis*manage) our money, for better or worse. We're confronted by self-control issues all the time, from the mundane—we procrastinate, waste hours on social media, have a third helping of dessert—to the dangerous and destructive—we don't take our medications, we have unprotected sex, we text and drive.

Coo-Coo for Choco Puffs

Why do we have such a hard time with self-control? It's because we tend to value certain things right now in the present much more highly than we value them in the future. Something that's great for us—but won't arrive for days, weeks, months, or years—isn't as valuable to us as something that's only okay for us but is available right now. The future simply doesn't tempt us as much as the present does.

In his famous marshmallow test, Walter Mischel left four- and five-year-old children alone, each with a single marshmallow. He told each child that if they did not touch the marshmallow for a short time, someone would bring them a *second* marshmallow—but only if they didn't touch the first one now. Most kids gobbled up their marshmallow right away, and never got to enjoy the second one.

But we're not kids, right? We're not impulsive; we have self-control. So answer this: Would you rather have *half* a box of delicious, designer, rare chocolates right now, or a *full* box of the same in one week's time? Imagine we passed the chocolate around so you could see it and smell it. It was right beneath your nose, right near your salivating mouth. What would you do?

Most people—most *adults*—say it's not worth it to wait another week for another half a box of chocolates, so they'll take the half box right now. So we're just like the little marshmallow-loving kids? Poop.

But wait! What if we push the choice to the future? Would we rather have the half box of chocolates in a year, or the full box of chocolates in a year and a week? It's the same question: Is it worth it to wait another week for another half box of candy? As it happens, when the question is presented this way, about the distant future, most people say they would prefer to wait another week for the larger chocolate box. In a year, it seems we believe waiting another week for an additional half box of chocolates is a worthwhile trade-off. Oh, so maybe we are adults after all!

Not really. The difference between our choice about now and our choice about the future is simply that decisions made in the present (some chocolate now or more in a week?) involve emotion, whereas decisions made about the future do not.

When we imagine our reality in the future—our lives, our choices, our environment—we think about things differently than we do in the present. Today our reality is clearly defined, with details, emotions, and so on. In the future, it is not. So, in the future we can be wonderful people. We will exercise, diet, and take our medication. We will wake early, save for retirement, and never text and drive. Imagine how enriched the world would be if everyone wrote the great American novels we've said we'll start "any day now."

The problem, of course, is that we never get to live in the future. We always live in the present. Today our emotions get in the way. Our emotions right now are real and tangible. Our emotions in the future are, at best, just a prediction. They are imaginary and, in our imagined future, we can control them. So this makes our decisions about the future emotion-free.

In the present, however, our emotions are real and powerful. They get us to succumb to temptation time and time again, and they cause us to make mistake after mistake. That's why every month—even those that were once "in the future" (pssst: they all were!)—Rob failed to save for retirement and gave in to buying a new speaker or a bottle of tire wax.

That's what happens when we add emotions to the decision-making mix: Now tempts us, but the future doesn't. Keeping our examples in the general area of the stomach, imagine we're asked which we'd prefer next month: a banana or chocolate cake? The banana is healthier, better for us. The chocolate cake is delicious. We'd say, "In the future, I'll take the banana." The future doesn't have any emotion, so the food choice just engenders a nutritional-value comparison. Which is better for us? But when we face the choice in the present, and pick between the banana and the chocolate cake, we think, "Right now, I really *want* the cake." In the present, we consider nutritional value and emotions, and desires, and wants. For most people, the chocolate cake creates much more emotional pull than the banana. To those for whom it does not, we apologize.

EMOTIONAL DEFINITION

Much of what makes us so emotionally detached from our future selves is the fact that our future selves are so poorly defined. We often imagine our future selves to be entirely different people than our present selves.[5] We understand and feel and connect to our current needs and desires much more than to our future ones.

The *immediate* rewards of one marshmallow or half a box of chocolates or better surround sound are vivid and salient, so they impact our decisions to a larger degree. The rewards of those things in the unknown future are much less salient, less tangible, less real to us, and because of that they make only a small dent on our decisions. An abstract future is harder to connect with emotionally than a real present.

Saving for the future—or the failure to do so—is a great example of the emotional difference between thinking about now versus later (when it comes to retirement, much, much later). When we save for retirement, we must give up something real right now for the enjoyment of our future self—and we have to make this sacrifice for a future self we can't quite connect to, a future self we often don't even want to think about. Who wants to think about being old and needy when we can be young and needy right now?

Since we should judge value based upon opportunity costs—what other things we could buy with the money we're about to spend—adding future spending into that equation makes considering opportunity costs even more complex. How do we compare the real temptation of buying tickets to see *Hamilton: The Musical* tonight against the possibility that the $200 ticket might be spent on some old-person medication we might need thirty years from now? It's very hard to do.

The issue of retirement savings is particularly complex and uncertain. We need to know when we'll stop working, what we'll be paid until that time, how long we'll live, what our expenses will be during retirement, and, of course, how our investments will pay off. Basically, who will we be, what will we need, what will the world provide for us, and at what cost in twenty, thirty, forty years? Easy peasy, right?

Tools for retirement planning aren't simple, either. There are plans and alternate plans and plans to manage the alternate plans while the management alters the alternate plans. There are tax concerns and defined benefits and defined contributions and the IRS and IRAs and 401(k)s and 403(b)s. Trying to figure it out can be intimidating and confusing. It's like trying to think of another word for "synonym" or what the best thing was before sliced bread. It's just tough to do.

To save requires us to value the distant, uncertain future and plan accordingly. That's something Rob couldn't do. It's something many of us can't do. Even if we can figure out the best way to save the most, we still face temptation and the challenges of self-control. It's easy to feel good now. It's hard to feel like we might not feel good later. As we've said already, and many others have said before, and we believe it's worth repeating: The benefit of consuming something now in the present will always outweigh the cost of passing it up to save for something else in the future. Or, as Oscar Wilde summarized the matter: "I can resist anything except temptation."[6]

Good Will Tempting

Most of us try to overcome temptation by applying willpower. But we rarely have enough of the latter to overcome the endless supply of the former. Temptation is everywhere, and with time and technology it is ever-increasing. Think of all the seemingly superfluous laws we need just to curb temptation—from preventing theft to keeping us from drinking and driving to controlling the abuse of painkillers to curbing intercousin marriage. There wouldn't be laws against these things if people weren't tempted to do them.

Consider, for a moment, texting and driving. Of course, we are capable of weighing the costs and benefits of reading a text immediately versus potentially crashing and dying or killing someone. No one ever said, "You know, I thought about the costs and benefits of checking my text while driving. I thought about the cost of taking a life. I thought about how much I want to stay alive myself. And I decided it was worth it to text! In fact, I am going to start texting more from now on."

No! Everyone recognizes that the moment we open a phone while driving increases the probability that we will die in a dramatic way. Everyone also recognizes that doing so is a really stupid way to risk our own lives and the lives of others. Nobody thinks it's a wise choice. Nevertheless, we keep on doing it.

Why are we so foolish? Because of these emotional factors—our inability to delay gratification, the uncertainty of dying from texting while driving, and our overconfidence in our ability to avoid death. Together these factors distort the value equation. We continue to be "perfect people" in the future, but that text is now. Now tempts us.

We spend more money than we know we should, eat more than we know we should, and, depending upon our divine affinity, sin more than we know we should. Temptation explains the gap between how we rationally know we *should* behave and how we emotionally do, whether with our wallet, our palate, or our pants.

When it comes to spending—and therefore not saving—the temptations are almost constant. We assume no one needs a primer on our culture of consumption, but just in case, turn on the TV, go online, read a magazine, or walk through a mall and feel the omnipresence of temptation.

Rob immersed himself in temptation. He surrounded himself with expensive entertainment equipment in his home and fancy bike gear on the road. These possessions constantly reminded him of what he has, who he is, and what he wants. Every month he *knew* he should save, but he couldn't overcome the temptation to spend. Like the kid in all of us—and the adult in most of us—Rob had very little self-control.

That's because self-control requires not just a recognition and understanding of the temptations of now, but also the willpower to avoid them. And willpower, by definition, requires effort—the effort to resist temptation, to refuse our instincts, to turn down a free marshmallow or fancy bike gear or anything that has any emotional resonance.

We don't fully understand willpower, but we know that it is a difficult power to harness.

Poor saving is really just one manifestation of poor willpower. But saving requires more than just willpower. To save we must first calculate a savings strategy, then we must acknowledge the emotions tempting us to veer away from that strategy, and then we must exhibit the willpower to overcome those temptations that await us behind every corner.

Obviously, it's easier *not* to start saving for retirement; this way we don't have to change any behaviors or reduce any of our present pleasures. It's easier to make some fatty microwavable snacks than to shop for, clean, and prepare fresh vegetables. It's easier to stay plump. It's also easier to justify our behaviors than to change them. It's not our fault that we occasionally sneak some of that chocolate cake: It's the chocolate cake's fault for being delicious.

Remote Controls

It's worth asking what other factors—besides discounting the future—reduce our willpower (which impacts our ability to overcome temptation . . . which uses our emotions to make us overvalue the present . . . which is why we have no self-control).

Everyone is aware of the human phenomenon of arousal. Some of us have even studied it, pretending to do so "for science." Dan, in fact, published a paper in 2006 with George Loewenstein that found that, when sexually aroused, men would do things that they would otherwise consider distasteful or immoral.[7] Another related paper found that men made poorer decisions while aroused. The paper was titled "Bikinis Instigate Generalized Impatience in Intertemporal Choice," because

"This Seems Like a Great Use of Research Funds and the Way I Want to Spend My Time" was too long.[8]

Besides arousal, other common factors that further increase our tendency to lose control include alcohol, fatigue, and distraction. Together, these make up the foundations of the casino and late-night infomercial industries. Mediocre music, mixed with the constant clinking of coins and dinging whir of slot machines, no visible doors or clocks, free cocktails, and pumped-in oxygen are the distraction tools of the casino. Rapid-fire edits, long-winded descriptions, and viewers' states of mind during the 3 a.m. programming block are the weapons of choice of late-night TV. Practitioners of each have built empires on the backs of our inability to resist temptation.

Working Together Against Ourselves

Of course, the problem of self-control doesn't work independently of the other valuation problems we've discussed. Rather, it amplifies those problems. We've spent all this time showing that it's really hard to think about money. It's challenging to weigh opportunity costs, avoid relative value, ignore the pain of paying, put aside our expectations, look beyond the language, and so on.

And now we're making things even more dire by explaining, in addition to all of those challenges, that a lot of financial decision-making is about the future. It's about what money, desires, and needs we will or won't have later and it is about the challenges of self-control. So, in addition to assessing the correct value of our current financial options, we have to think about the future, which makes things extra hard.

Remember Brian Wansink (*Mindless Eating*) and his bottomless bowl of soup in our discussion of relativity? Well, people did not keep eating the soup only because of the hunger cues caused by relativity (the quantity of soup as judged by the size of the bowl). That is, we often eat just because we see food—not because we're hungry, but because it's there. It's our instinct to eat because eating *feels good.* It's tempting, it is immediate, it is now. Without self-control, there's nothing to stop us but the retreating bottom of a bottomless bowl.

At least we're not fish. If we put too much food in a goldfish tank, our goldfish—let's call her Wanda—will keep eating until her stomach explodes. Why? Because fish have no self-control. And Wanda didn't read this book. So when we feel down about our self-control, remember the fish. Compare ourselves to Wanda, and feel good. Relatively good.

The pain of paying carries some implications for self-control. The pain of paying makes us aware of our choices. It makes them more salient and helps us master some self-control. If we use cash instead of a credit card, we're more likely to feel the impact of a sudden $150 dinner with friends. That feeling in the present helps us fight off the temptation of the expensive meal. In the same way, mechanisms that diminish the pain of paying help us short-circuit self-control and get us to fall for temptation more easily and quickly.

Mental accounting—especially malleable mental accounting—is another tactic we use to weaken our self-control. "I shouldn't go out to eat tonight—but what if I call it a work event? Yum!"

When we discussed overtrusting ourselves, we focused on trusting our *past* selves—either the self that made a money decision in the past or the self that saw an irrelevant price, like a real estate listing price. But we also have trust issues between our current selves and our future selves. Rob's future self trusts his present-day self to forgo immediate gratification to save for retirement, while his present self trusts his future self to make smarter, more selfless decisions about . . . saving for retirement. Neither has proven trustworthy. For the rest of us, relying upon our future or past selves to resist or have resisted temptation is equally unwise.

These forces and the other issues we've discussed cause us to assess value incorrectly. Our lack of self-control, however, makes us act irrationally whether we value things correctly or not. We might think we've navigated all the psychological pitfalls to come to a rational financial assessment . . . but then, in many cases, our lack of self-control makes us do the irrational thing anyway. The struggle to maintain self-control is like facing a luxurious dessert cart after struggling through a kale and quinoa dinner. Come on, you only live, and spend, once. Right?

NOT-SO-EASY MONEY

Dan once attended a conference with luminaries from the world of sports. Muhammad Ali was there, and, of course, it was hard not to think about the long-term impact of his boxing career on his life. Ali was willing to endure brutality for the success of a boxing career, only to pay for it later with the effects of Parkinson's disease. We will not judge his decisions—we don't know what factors he considered or what science was available to him at the time or what else informed his choices—but in Ali's life one can easily see the disconnect between our present desires and future well-being.

At the same conference there was also a well-known baseball player who told Dan about signing his first professional contract. When his coach gave him his first paycheck, to his shock there was only $2,000 in it. He had signed a contract for millions of dollars, so he didn't understand why he was getting so little.

He called his agent, who told him, "Don't worry, I have your money. It's safe. I am going to invest it for you so that when you retire you will not have to worry about anything. In the meantime, I have given you spending money. If you think you need more to live on, just let me know and we'll talk about it."

The player's peers were making similar giant salaries, but they didn't have the same agent. So they were spending more, driving nicer cars, and doing more expensive things. But they weren't saving nearly as much as he was. Now, years later, most of them are broke, while this player and his wife are living well, thanks to a lifetime of saving.

This ball player shined a light on a surprising set of facts. Many professional athletes make a lot of money quickly. They also spend a lot of money in a short time and very often declare bankruptcy quickly. About 16 percent of NFL players file for bankruptcy within twelve years of retirement, despite average career earnings of about $3.2 million.[9] Some studies say the number of NFL players "under financial stress" is much higher—as high as 78 percent—within a few years of retirement. Similarly, about 60 percent of NBA basketball players are in financial trouble within five years of leaving the game.[10] There are similar stories about lottery winners losing it all. Despite their big paydays, about 70 percent of lottery winners go broke within three years.[11]

Earning or winning a great deal of money intensifies the challenges of self-control. Oftentimes, a sudden increase in wealth is particularly challenging. Counterintuitively, adding a whopping sum to our bank accounts is no guarantee that we will be able to better manage our finances.

Jeff has a hypothesis he would very much like to study: He believes that, unlike most people, he would be able to manage a sudden influx of cash. Sadly, he has been unable to secure the appropriate seven-digit funding for this project, but he holds out hope that someone will support this important scientific research soon.

Almost everything about our culture encourages and rewards the loss of self-control. "Reality" television is all about who behaves the worst—who loses it, who acts out, who goes nuts. They don't air "Do You Eat Vegetables Better Than a 5th Grader." The TV series *Temptation Island* wasn't about the formation of the beloved group the Temptations, and *Here Comes Honey Boo-Boo* was not about a responsible-but-clumsy beekeeper.

Self-control problems are everywhere. They have been with us forever, from the time of Adam and Eve and that ripe, juicy apple (or whatever our original sin of choice may be).

Not only is temptation everywhere, but it's getting worse. Think about it: What does the commercial environment around us want us to do? Does it care about what is good for us in twenty or thirty years? About our health, family, neighbors, productivity, happiness, or waistlines? Not likely.

Commercial interests want us to do whatever is good for them and to do it right now. Stores, apps, websites, and social media clamor for our attention, time, and money in ways that are good for them in the short term and without much (or any) concern for our long-term best interests. And guess what? They know, typically better than we do, how to push our buttons. And they get better at this all the time.

As a consequence of this increasing temptation, the really bad news is that we have many self-control problems and we'll likely wind up with many more. As phones, apps, TVs, websites, retail stores, and whatever the next commercial frontier is get better, they also get better at tempting us. The good news: We're not helpless. We can overcome some of these problems by learning about our behavior, about the challenges we face, and about how our financial environment encourages us to make poor choices. And we can use technology to help us overcome—to help us think about using money to serve our own long-term interests, rather than serving others'.

More on that in a bit. Can you wait? Do you have the willpower to fight off the temptation to skip ahead for some solutions? We think you do.

13

WE OVEREMPHASIZE MONEY

Way back around the turn of the century—that is, around the year 2000—a young(er) Dan Ariely was looking to buy a couch for his office at the Massachusetts Institute of Technology. His search led him to a fine sofa that cost $200. Shortly thereafter, he found another sofa by a French designer that cost $2,000. It was much more interesting, very low to the ground, and sitting in it felt very different. But it wasn't clear that it was more comfortable or that it would serve its role as a sofa in a better way. It certainly didn't seem to be worth paying ten times as much. But Dan bought the fancy one anyway. Since then, when guests of all sorts come to his office, they have a hard time lowering themselves onto this sofa and an even harder time getting out of it. We will not address rumors that Dan has kept this sofa simply for the purpose of torturing his visitors.

What's Going On Here?

Dan had a hard time evaluating the long-term experience offered by the fancy sofa. He tried it out by sitting in it for a few minutes, but the real questions were how comfortable the sofa would be after sitting in it for more than an hour—which turned out to be very comfortable—and how his guests would feel using it—which turned out to be not so great. (After many years, Dan now knows that certain guests don't feel comfortable sitting so low and that they have a hard time getting up again.) Without having a way to answer these questions at the time of purchase, and thus a way to know how suitable the sofa would be for his needs, Dan used a simple heuristic: Expensive must mean good. So he got the expensive couch.

Dan is not alone in using this decision strategy. Would you eat cheap lobster? What about discount caviar or bargain-basement foie gras? Restaurants don't put delicacies like these on sale because of how we deal with the price, and the powerful signals it sends. Even if the wholesale markets for lobster, foie gras, and caviar plummeted, as happened a few summers ago, restaurants won't pass those savings on to diners. That's not just because they're greedy, but because low prices send us uncomfortable messages about the nature of luxury items. We infer that discounts mean lower quality. We start thinking there's something wrong with the weird little food things. We certainly assume they're inferior to competitors' delicacies.

What if, instead of cheap lobster and foie gras, we were offered extremely inexpensive heart surgery? Same thing: We would think something's wrong and would seek out the best surgeon we could find, which, given our lack of knowledge about cardiology, would probably be the most expensive one we could find.

That's because another important way we value things—a way unrelated to actual value—is by assigning meaning to a price. When we can't evaluate something directly, as is often the case, we associate price with value. This is especially true in the absence of other clear value cues. Dan, as a young, impressionable MIT professor, didn't know how to measure the value of an office couch, so he went with what he could measure: price. A decade and a half and many unhappy guests later, he knows he made a poor choice.

In *Predictably Irrational*, Dan showed that we are conditioned to see high price as a stand-in for effectiveness. Dan, along with his colleagues Rebecca Waber, Baba Shiv, and Ziv Carmon, did an experiment with a fake painkilling drug they called VeladoneRx.[1] (In truth, it was a vitamin C capsule.) They gave it to test subjects along with brochures and a technician in a crisp business suit and white coat and slapped on an expensive price tag of $2.50 per pill. They then gave participants a set of electrical shocks to see how much pain they could take. Almost all the patients in the study showed reduced pain after ingesting VeladoneRx. When Dan and his partners in crime carried out the same experiment, using a price tag of 10 cents per pill, the average amount of pain relief patients experienced was about half of that under the $2.50 pill.

Baba, Ziv, and Dan extended these findings using Sobe energy drinks. In these experiments, as mentioned earlier, those who had the beverage along with literature claiming it improved performance actually displayed improved performance on all kinds of mental tasks. Another experiment showed that those who received discount-priced energy drinks performed worse than those who drank full-priced beverages. Another experiment showed that those who got the discounted drinks expected them to be worse, and indeed they experienced them as worse because of the signals sent by price.[2]

Whether it makes sense or not, a high price signals a high value. In the case of important things like health care, food, and clothing, it also signals that the product isn't cheap or of low quality. Sometimes the absence of poor quality is as important as the presence of high quality. Aunt Susan may not pay $100 for a T-shirt, but if that's JCPenney's "regular" price, then, the rationale goes, someone must be willing to pay it. Therefore, it must be a high-quality product. And lucky Aunt Susan, she just got one of those fancy $100 T-shirts for $60. The Vertu cell phone offers the same service and functionality as most other phones, but those who can afford it pay between $10,000 and $20,000 for the honor of playing Angry Birds on a prestigious status symbol. "Surely no one would pay that much if it wasn't worth it," someone must have reasoned and then went ahead and got a Vertu. On another technology platform for only one day—because it was quickly taken down—there was an iPhone app for sale called "I Am Rich." It simply displayed a few words of affirmation about being rich. It did nothing else. It cost $999.99. Eight people bought it. We would like to invite those eight people to contact us about some other similarly promising opportunities.

Prices shouldn't affect value, performance, or pleasure—but they do. We are trained to make quick decisions based on money with every single transaction, and, especially in the absence of other value markers, that's what we do.

Remember that anchoring and arbitrary coherence show that just listing a price can impact our perception of value. (The first price we see associated with a product anchors our valuation of it, and it doesn't even need to be a price; it can be an arbitrary number like a Social Security number or the number of countries in Africa.)

Consider wine, the best way to a man's stomach, which, as we've heard, is then the way to his heart. The higher the price of a bottle of wine, the more we like it. The evidence is clear: When we know how much we're spending on what we're drinking, then the correlation between price and enjoyment is incredibly strong. And it doesn't matter much what the wine is.[3] However, using price to infer quality is a fairly blunt assessment. The impact of the price on this inferred quality might be reduced if we could judge the wine in other ways—if we know where the wine is from, when it was grown, why that matters, or if we know the winemaker personally and how he or she washes his or her feet before crushing each individual grape. But that seems unlikely.

Uncertain Situations

That's all well and good, but how often do we "know the winemaker"? That is, how often do we know all the relevant details that would allow us to objectively assess the value of a safari or a widget or a safari full of widgets? Hardly ever. As we've seen, we usually don't have any idea what anything should cost. Without context, we have no independent ability to truly value anything, be it casino chips, home prices, or Tylenol. We are afloat in a sea of financial-value uncertainty.

In times like this, money becomes the salient dimension. It's a number. It's clear; we can compare it across multiple options; and because it's easy to think about money in this literal, seemingly precise way, we pay too much attention to it at the expense of other considerations.

Why is this? Well, it's about our love of precision. There's a saying that with regard to our decisions in general, and our financial decisions in particular, psychology gives you a vaguely right answer and economics gives you a precisely wrong answer.

We love precision—and the illusion of precision—because it gives us the feeling that we know what we are doing. Especially when we don't.

The strange thing about money is that, even though we don't understand what it is, it's measurable. Whenever we encounter a product or experience with many different properties, along with one precise and comparable attribute (money), we tend to overemphasize that specific attribute because it's easier to do so. It's hard to measure and compare features like flavor, style, or desirability. So we end up focusing on price as a way to make our decisions, because we can measure and compare it more easily.

People often say they'd prefer being the highest-paid employee of a company rather than the lowest-paid one—even if it means making less money. Ask people if they'd prefer to make $85,000 and be top dog or earn $90,000 and not be, and they'll choose the $90,000. Make sense? Yes.

But if we ask the same question with a different focus, we get a very different answer. When we ask people if they would be *happier* if they made $85,000 and were the highest paid or if they made $90,000 and were not—the same options with the same parameters, just framed in terms of happiness—they say they would be happiest making only $85,000. The difference between how people respond to the problem in general versus when focused on happiness is due to the fact that it is very easy to think just about money. In the absence of another specific focus, money is the default focus. When we think about something like a job, even though there are many things that come into play, money is so specific, precise, and measurable that it comes to mind most quickly and plays a large role in our decision.

To consider a more mundane example of the same principle, consider our nightmare of choosing a cell phone. There are many factors—screen size, speed, weight, camera pixels, security, data, coverage. Given all these factors, how much weight should you give the price? Well, as a product's complexity increases, relying on the price becomes a relatively simpler and more attractive strategy, so we focus on the price and largely ignore the many complexities of that decision.

Along the same lines, as we learned in the discussion of arbitrary coherence, most of us have a hard time comparing one type of product or experience to a very different one. That is, we don't use opportunity costs to compare a Toyota to a vacation or to twenty expensive dinners. Instead, we compare things in the same category—cars to cars, phones to phones, computers to computers, widgets to widgets. Imagine we bought the first iPhone, which was the only smartphone at the time. There was no similar product to compare it to, so what would we compare it to? (Yes, Palm Pilot and BlackBerry were around back then, but the iPhone was so far ahead as to be a completely different category of product. Also, Palm Pilot? No thanks, Grandpa). How would we figure out if it was worth the cost? When Apple first introduced the iPhone, the price was $600. A few weeks later, the company reduced the price to $400. That created a new category to which to compare the iPhone—the *first* iPhone, which was, in fact, the identical iPhone at a different price. Once there are multiple products in a category, money becomes an alluring way of comparing them, which can in turn lead us to overemphasize price. We focus on the price difference (wow, it is $200 cheaper) rather than on other qualities, and of course we continue to ignore opportunity costs.

Money is not the only attribute that is easily used for points of comparisons. Other attributes, if we quantify them, can also function in the same way. But these same attributes—if we don't quantify them—are much too difficult to use. It's hard to measure the deliciousness of chocolate or the drivability of a sports car. This difficulty shows the gravitational pull of price: It is *always* easy to quantify, measure, and compare. For instance, megapixels, horsepower, or megahertz, once specified and held up side by side, become more comparable and precise. This is called *EVALUABILITY.* When we compare products, features that are quantifiable become easy to evaluate, and even if they are not truly important they nevertheless come into sharper focus, which makes it easier for us to evaluate our options in terms of those features. Often these are the features that the manufacturer wants us to focus on to the exclusion of others (in other words, let's talk about pixels, not how often this camera breaks). Once an attribute is measured, we pay more attention to it and its importance on our decision grows.

Christopher Hsee, George Loewenstein, Sally Blount, and Max H. Bazerman once ran an experiment in which they asked people browsing used textbooks how much they would pay for a music dictionary that had 10,000 words and was in perfect condition. Another group was asked how much they would pay for a music dictionary with 20,000 words but a torn front cover. Neither group knew about the other dictionary. On average, the students were willing to pay $24 for the 10,000-word dictionary and $20 for the cover-torn 20,000-word one. The cover—irrelevant to looking up words—made a big difference.

The researchers then cornered another group and presented them with both options simultaneously. Now the students could compare the two options side by side. That changed their perception of the products. In this easy-to-compare group, the students said they would pay $19 for the 10,000-word dictionary and $27 for the 20,000-word one with the torn cover. Suddenly, with the introduction of a more clearly comparable aspect—number of words—the larger dictionary became more valuable, despite the torn cover. When evaluating only a single product, the study participants weren't sensitive to whether the dictionary had 10,000 or 20,000 words. It was only once that attribute was easily comparable that it became an important factor in assessing value. Again, when we don't know how to evaluate items, we are disproportionally affected by features that are easily comparable, even when those features (the torn cover, in this instance) have little to do with the real value of the product in question. In this case, the importance of the number of words increased and the importance of the condition of the cover dropped. More often than not, though, the feature we overemphasize when we make our decisions is the one thing that is always easy to see and evaluate: price.[4]

So, if we tend to focus on whatever is most measurable and comparable, is there something wrong with that? Well, yes. It can be a big problem when the measurable thing is not the most important part of the decision. When it is not the desired end, but just the means to that end. A good example is frequent-flyer miles. No one's life aspiration starts and ends with the accumulation of frequent-flyer miles—they're merely a means that can one day procure the desired end of a vacation or free flights. Even George Clooney's character in *Up in the Air* strives to gather miles not for themselves, but for other reasons, as a symbol of power and prosperity.

While few people consider maximizing frequent-flyer miles to be the key to a life worth living, it's tempting to maximize anything that's easily measurable. How do we compare 10,000 more miles with four more hours of relaxation on the beach? How many miles equals an hour of relaxing?

Money works the same way. It isn't the final goal in life, it's a means to an end. But because money is much more tangible than happiness, well-being, and purpose, we tend to focus our decision-making on money instead of on our ultimate, more meaningful goals.

We want to be happy and healthy and enjoy our lives. Measurable things like frequent-flyer miles and money and Emmy nominations are among the easiest ways to gauge our progress. People will often choose to fly crazy routes just to get more miles, the process of which actually reduces their overall happiness due to flight delays, uncomfortable seats, and the talkative sales guy who won't shut up about his crush on Mavis from the copy center. Just ask her out, already!

Winning the Game of Life

Ah yes, life. And money. And what is important.

Money is a signifier of value and worth, which is, for the most part, a good thing. Our lives are individually and collectively more vibrant, enriched, and free because of money. But it's not so good when money's role as a measure of value and worth extends into parts of our lives beyond goods and services.

Since money is more tangible than human needs like love and happiness and a child's laughter, we often focus on money as an approximation of our lives' value. When we stop to think about it, we know that money isn't the most important thing in life. No one ever lies on their deathbed wishing they'd spent more time with their money. But because money is much easier to measure—and less frightening to consider—then whatever the meaning of life might be, we can focus on it instead.

Consider how an artist's work is valued in a modern economy that doesn't pay for content creation as it once did. Since money is how our culture defines value, not getting paid for your work can be both insulting and demoralizing, even though money is, arguably, not the goal of art. Many of the great artists of history either relied upon generous patrons, the likes of whom do not exist anymore, or died destitute . . . and that was back when they didn't have to compete for attention with Candy Crush and Instagram models.

Throughout Jeff's nontraditional career—lawyer for about three minutes, comedian, columnist, author, speaker, male underwear model (not really, but one can dream)—his family greeted every one of his accomplishments, from writing a book, to getting on TV, to making connections, to meeting Dan (it was through Jeff's first book on dishonesty, not on Tinder, as the rumors may have it), with the question "What does it pay?" For a long time, this bothered him, because it seemed callous and dismissive, a clue that they didn't understand the true value of what he was doing. Well, they didn't understand what he was doing, but they were not dismissive. They were trying to understand. They were using the money question as a proxy in an attempt to learn. Seeking monetary terms was a bridge for them to reach out, to translate the intangible, incomprehensible steps Jeff was taking into a language they could understand: money. At first, that was a painful difference between how Jeff and the people around him saw the world, but as Jeff realized that it was not just criticism but also an attempt to understand, it became a bridge of common language. It helped them analyze what he was doing and attach judgments and values and advice and support. This way they could ridicule his choices with *informed* put-downs, *reality-based* jokes, and *educated* eye rolls. Progress.

Of course, while some focus on money is understandable, some might say we all left the useful parts of that focus behind long ago and are now aimlessly powering through the seas of financial uncertainty wholly obsessed with money.

Apples to Apples, Dust to Dust

We should realize that money is just a medium of exchange. It allows us to exchange things like apples and wine and labor and vacations and education and housing. We shouldn't attach symbolism to it. We should treat it as what it is: a mere tool to get us what we need, want, and desire, now, a bit later, and much later than that, too.

There's the old expression about how difficult it is to compare apples to oranges. But that's not true. Comparing apples and oranges is actually very easy: No one ever stands by the fruit plate wondering if they prefer the apple or the orange. When we value things by how much pleasure they would give us—what's known as a direct hedonic evaluation—we know with high certainty which option is expected to give us more pleasure.

What's hard is comparing apples to money. When we bring money into the equation, we make the decisions much more difficult and we open ourselves up to mistakes. Determining how much money is equal to the pleasure we expect to get from an apple is a calculation fraught with danger.

From this perspective, a useful financial decision-making strategy is to pretend that money doesn't exist.

What if we took money out of the equation from time to time? What if, instead of looking at a vacation, we quantified the amount that this vacation would cost us in terms of movies we could attend or wine we could drink? What if we looked at the wardrobe we were going to replace for the winter and we calculated how many tanks of gas or bicycle repairs or days off work it would cost? What if, rather than considering the difference in price between big-screen TVs, we were to think about the difference as a dinner out with friends and fourteen hours of overtime and then decide if that's worth the upgrade or not?

When we move from comparing *money* to things to comparing *things* to things directly, it puts our choices into new perspective.

This process may be most applicable and useful for big decisions. Imagine we have the option to buy a big house and spend a lot on a mortgage, or a medium house with a smaller mortgage. It's hard to compare these options when the terms are in dollars a month and a down payment and interest rates and the like. The decision gets even harder when everyone involved in the process—the sellers, the agents, the mortgage lenders—wants us to spend more to buy the larger house. What if we didn't think in terms of money? What if we said, "You know what, the bigger house costs me the same as the smaller house plus one yearly vacation, a semester of college for each of my children, and an additional three years of working before retirement. Yes, I can afford it, but maybe it's not worth exchanging all those things for an extra bathroom and a larger yard." Or maybe we do that calculation and still decide the bigger house is worth it. Great! But at least we are making a clear-eyed decision by considering some alternative ways of using our money.

This direct-comparison method is not necessarily the most efficient, or even the most rational, approach. It would be crippling to take the time to translate every transaction into a money-free opportunity cost analysis. But it is a good exercise with which to assess our decision-making abilities, particularly when we face large decisions.

Money is a curse and a blessing. It's a wonderful thing to have money as a medium for exchange, but, as we've learned, it often misdirects us and influences us to focus on the wrong things. For an antidote, a bit of moneyless reframing helps from time to time. Consider the underlying trade-offs between things and other things instead of between things and money. If you're happy with the trade-off, go for it. If you're not, think again. And again. And again.

No matter our station in life, we believe it is important that instead of thinking about life decisions in terms of money, we think about them in terms of life.

Money in Charge

You may remember one or more of the people we met in the chapters of this book: George Jones, Aunt Susan, Jane Martin, honeymooning Jeff, the Tucson Realtors, Tom and Rachel Bradley, James Nolan, Cheryl King, Vinny del Rey Ray, and Rob Mansfield. They spent a lot of time trying to figure out how to spend their money, yet they still got it wrong. They were fools, not just because they couldn't figure out the complex and convoluted world of money, not just because they fell for irrelevant value cues, not just because they made mistakes, but also because they spent so much time worrying about money. They were afloat on that sea of uncertainty, and allowed themselves to be moved along by value cues that deposited them, like ritual sacrifices, onto the base of a money volcano.

This chapter started with an analysis of how we all overemphasize money—specifically price—when we try to assess value in our financial decisions. It then analyzed how we might overemphasize money in other important decisions and in evaluating our lives in general.

Neither of us is competent or qualified or 110 percent blissfully happy enough to tell anyone what to do with their lives, but we have sufficient data to show that we should aim to be more free from the overbearing burden of money. Or at least have it loosen its grip on us a little bit.

We don't want to tell you how to prioritize things, where you should place money on the sliding scale of family, love, good wine, sports teams, and naps. We just want you to think about how you think about money.

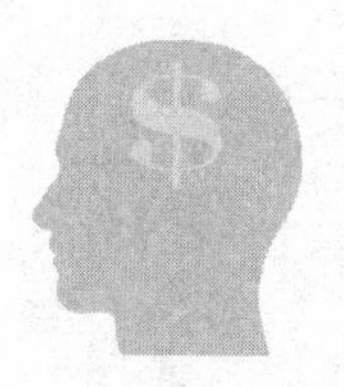

PART III

NOW WHAT? BUILDING ON THE SHOULDERS OF FLAWED THINKING

14

PUT YOUR MONEY WHERE YOUR MIND IS

So now what?

We've seen how we think about money incorrectly, how we assess value in ways that have little to do with actual value, and how these get us to misthink and misspend our money. We've gotten a peek behind the curtain—a glimpse at the inner workings of our financial brain. What we've learned is that we overemphasize irrelevant factors, forget about important ones, and allow insignificant value cues to lead us astray.

So how should we think about money? What are the solutions to all our problems?

We're sure some of you have just flipped to the back of this book to find out. Many of you may be doing so while simply browsing at the bookstore. If so, we 1) applaud you for saving the cost of this book, but 2) suggest that you're not correctly valuing our effort, and 3) offer here the short version: When it comes to making financial decisions, what *should* matter are opportunity costs, the true benefit a purchase provides, and the real pleasure we receive from it compared to other ways we could spend our money.

What should *not* matter in a perfectly rational world?

- Sale prices or "savings," or how much we're spending at the same time on something else (relativity)
- The classification of our money, where it came from, and how we feel about it (mental accounting)
- The ease of payment (pain of paying)
- The first price we see or previous prices we've paid for a purchase (anchoring)
- Our sense of ownership (endowment effect and loss aversion)
- Whether someone appears to have worked hard (fairness and effort)
- Whether we give in to the temptations of the present (self-control)
- The ease of comparing the price of a product, experience, or widget (overemphasizing money)

Remember: Those things do not affect the value of a purchase (even if we think they do). There are other factors that would not change value if we were perfectly rational, but since we are full of quirks, they end up changing the value of our experiences. These include the following:

- The words describing something and what we do at the time of consumption (language and rituals)
- How we anticipate the consumption experience, rather than what its true nature is (expectations)

Language, rituals, and expectations are in a different group from the other factors because they can change the experience. A 25 percent discount or one-click payment will never change the value of an item. Learning about the winemaking process and having a white-gloved sommelier pour a glass at a lakeside picnic can make the whole experience more meaningful, interesting, and valuable.

If we were perfectly rational, language, rituals, and expectations should not influence our spending decisions. However, because we are humans and not robots, it's hard to say that language, rituals, and expectations should *never* influence us. It's hard to say when taking these forces into account becomes a mistake, especially when they provide an enhanced experience. If we expect to get more from a wine—because of the descriptive language, setting, bottle, tasting rituals, and so on—we *will* get more from the wine. So, is allowing that to happen a mistake? Or is this an added value for which we should be willing to pay?

Whether or not language, rituals, and expectations are welcome additions to any particular valuation, what is clear is that we should be the ones making that decision to add them or not. We should be the ones choosing to dive deeper into these irrationalities in order to get more value, rather than having those influences forced on us. With the awareness we now have, we can decide if and when to enjoy wine more just because of how it's poured.

Frankly, we're not sure we want to live in a world without language, rituals, and expectations—a world in which we'd therefore experience things in purely neutral emotional states. That doesn't sound like fun. We just want to ensure we're in control of the ways that these important elements are used.

There, that's simple. From relativity to expectations, now you know how we think about money and the irrational biases that affect us when we do. Now go make every financial decision with all those lessons in mind.

Not so easy, right? Seems pretty daunting. Well, there's a reason why we've decided to show you *why* we make foolish money decisions, rather than telling you *what to do* in any situation. For one, we just don't know what is the right thing to do in every situation. No one knows. But we also don't want to give you fish; we want to show you how you've been fishing, so you may approach future fishing in a better way, if you so choose. Maybe that's not fair—to dump a bunch of information on you and bid adieu. To point out that we are up the creek without a paddle and then swim away. To say, "We're doomed," and then laugh.

Except we actually don't think we're doomed. We're actually optimistic. We believe that we have it in ourselves to overcome many of our money mistakes.

If we put our minds to it, we can individually and collectively improve our financial decision-making. The first step is being aware, and we've achieved that. The next step is turning that awareness into an effective plan, into concrete steps, into change.

Now that we've studied the many things we do badly, we can begin to examine the nuances of our behavior in order to find tools that will help us to build a better future. One of the main lessons of behavioral economics is that small changes to the environment we live in matter. Following this approach, we believe that a detailed understanding of human frailty is the best first step toward improving the ways we make decisions in general and financial decisions in particular.

Let's start by considering what we can do, individually, to avoid, correct, or mitigate each of the valuation mistakes we make.

We ignore opportunity costs

Think about transactions in terms of opportunity costs by considering more explicitly what we're sacrificing for what we're getting. For instance, we can translate dollars into time—how many hours of wages, or months of salary, we must work to pay for something.

We forget that everything is relative

When we see a sale, we shouldn't consider what the price *used to be* or how much we're saving. Rather, we should consider what we're actually going to spend. Buying a $60 shirt marked down from $100 isn't "saving $40"; it is "spending $60." Aunt Susan never actually got $40 in her pocket, but she did get an ugly shirt on her back. Or, more likely, her nephew's back.

When it comes to large, complex purchases, we can try to segregate our spending. That is, when we buy something with many options—like a car or a house—we should judge each additional item separately.

We should try not to think in percentages. When the data is presented to us in percentages (for example, 1 percent of assets under management), we should do the extra work and figure out how much money is really on the line. The money in our pocket is tangible; it exists in absolutes—$100 is $100. Whether it's 10 percent of a $1,000 purchase or 1 percent of a $100,000 one, it still buys the same 100 packs of Tic Tacs.

We compartmentalize

Budgeting can be useful, but remember this simple principle: Money is fungible. Every dollar is the same. It doesn't matter where money comes from—our job, an inheritance, a lottery ticket, a bank robbery, or our gig moonlighting as the bassist in a jazz quartet (dare to dream)—the money is all ours and it belongs, in fact, to the general "our money" account. If we find ourselves splurging with certain "kinds" of money—just because in our mind the money belongs to the "bonuses" or "winnings" account—we need to pause, think, and remind ourselves that it's just money. Our money.

At the same time, we should remember that using mental accounting to categorize our *spending* can be a useful budgeting tool for those of us who can't do constant, instantaneous opportunity cost calculations. That is, all of us. It is a potentially dangerous tool because, on one hand, it opens us up to inconsistencies in how we use money. But, on the other hand, if used correctly, it can help us stay in the general vicinity of the ways we want to spend our money.

We avoid pain

The pain of paying may be the trickiest, and most ominous, of all the ways we mess up with money. Maintaining some pain of paying helps us at least consider the value of our options and the opportunity costs that lie within. The pain helps us pause before purchasing and consider whether or not we really should spend our money then and there—it helps us consider opportunity costs.

The problem, of course, is that the people who make payment systems don't share our desire to slow down, consider alternatives, and think. This is why the best solution for the pain of paying may be as simple as "Don't use credit cards." Or maybe it's an even simpler "Punch yourself every time you spend money so you really feel it." That might not be a sustainable financial plan, though, since eventually the medical bills will catch up with us.

Realistically, we won't suddenly stop using credit cards. But we should be skeptical of the latest financial technologies, especially those that are designed to demand less of our time and attention and make it easier for us to part with our money. It won't be long before blinking in a certain way will be a payment option. Don't sign up for that.

We trust ourselves

Trusting ourselves—our past judgments, choices, and responses to prices we've encountered—is normally considered a good thing. "Trust your gut," the self-help gurus yell (for a hefty fee). That's often not a good idea, particularly in the context of spending. When it comes to spending, trusting our past decisions contributes to the problems of anchoring, herding, and arbitrary coherence. So we should question seemingly "random" numbers, prominently placed MSRPs, and insanely high-priced products. When we see a $2,000 shoe or a $150 sandwich, watch out for the second-most-expensive shoe or sandwich or shoe that somehow doubles as a sandwich.

In addition to questioning the prices others set, we should also question the prices we set ourselves. We should avoid doing something all the time, like getting a $4 latte, just because we've always done it before. From time to time, let's stop and question our long-term habits. Those of us who do not learn from our own spending histories are doomed to repeat them. We should ask if a latte is really worth $4 to us, or if a cable bundle is worth $140 per month, or if a gym membership is worth fighting for parking just to look at our phone while trudging on a treadmill for an hour.

We overvalue what we own and what we might lose

We shouldn't trust that the home renovations we are going to make will increase the resale value of our home. We should recognize that our taste is unique, and that other people might see things differently. Renovating is fine, as long as we head down that path recognizing that it might only increase the value of the home to us.

We should watch out for trial offers and promotions. Marketers know that once we own something, we will value it more and have a harder time giving it up.

Sunk costs cannot be recovered. If an amount is spent, it's spent. The past is past. When making decisions, consider only where we are now and where we will be in the future. We may think sunk costs should affect future decisions, but they don't. We need to do what millions of four-year-old *Frozen* fans have screamed into their parents' faces the last few years: "Let it go! Let it gooooo!"

We worry about fairness and effort

There's a simple lesson that we all learn at some point in life, whether it's as a five-year-old who gets pushed off a swing or a thirty-five-year-old who is passed over for a promotion: The world isn't fair. Sorry.

Let's not get caught up in whether something is priced fairly; instead, consider what it's worth to us. We shouldn't pass up great value—access to our home, a salvaged computer, getting a ride in winter weather—just to punish the provider for what we think is unfairness. They probably won't learn the lesson, and we'll be stuck outside in the rain and snow with no computer files.

We may also be wrong about whether something is a fair price, and about whether or not it took a lot of effort. Let's also recognize that there is value in knowledge and experience. Locksmiths, artists, authors of books about money—the value of their work does not come from the time and effort we witness, but from the time and effort they've spent developing their expertise over a lifetime. Craftspeople have perfected the art of making what they do look effortless, but it's not. From Picasso to parenting, sometimes the most difficult jobs look easier than they really are.

But let's be careful not to fall for false effort. We ought to watch out for too much transparency. If a consultant shows us all the great pains they have gone through to produce nothing but their $100,000 fee, reconsider. If a Web page is just a progress bar and a "Pay Now" button, keep searching. If our spouse grunts and groans, wails and screams, feigns agony and despair while loading the dishwasher or doing the laundry—well, in that case, we should probably offer them a foot rub. Just to be safe.

We believe in the magic of language and rituals

The great twentieth-century philosophers Public Enemy (they're also a hip-hop group) put it best: "Don't believe the hype." If the description of something, or the process of consuming something, is long-winded and overblown, we're probably paying for that description and process, even if it doesn't add any real value.

Watch out for irrelevant effort heuristics: There is rarely reason to pay for an artisanal hammer.

At the same time, remember that language and rituals can change the quality of our experiences, so we should embrace them to enhance experiences if we so choose.

We make expectations a reality

Expectations give us reason to believe that something will be good—or bad, or delicious, or gross—and they change our perception and experience without altering the true underlying nature of the thing itself. We should be aware of the source of expectations—whether it's the pleasure of dreams and aspirations or the irrelevant allure of brand names, biases, and presentation. Or, as many great philosophers and mediocre graphic designers have put it, "Don't judge a book by its cover."

As with language and ritual, we—Dan and Jeff—want to acknowledge, again, that expectations actually can alter our experiences. We can *use* such expectations to our advantage or they can *be used* by others to take advantage of us.

Once we buy a bottle of wine, we may want to manipulate ourselves into believing it's worth $20 more than we paid. We can let it breathe and swirl it and smell it and put it in a fancy glass knowing that with all of these tricks, it's going to be a better experience. That's using expectations.

What we don't want is to buy a bottle of wine because someone has tricked us into spending $20 more than we should. We hear the sommelier describe the vintage and tannins and awards and labels and reviews and hints of elderberry and believe it must be worth a lot. That's being used *by* expectations.

What is reality? Is it the objective taste of wine as a robot would taste it, or does the taste include our expectations and all the psychological influences around it? In truth, both are realities. Imagine there are two bottles of the same wine, but one has a different shape, color, label, and recommendation. Our expectations could make us experience those two bottles very differently. A blind taste test—or a taste test by a robot—would find that each bottle tastes the same.

But we don't live life as blind robots. (Well, we don't know everything happening with artificial intelligence and neuroscience, so maybe we do, but most of us remain human.) We shouldn't discount the reality where our expectations can objectively improve our enjoyment of a wine. That happens. That is also real.

It's a choice of manipulation versus self-manipulation. We don't want to be manipulated unwillingly or unconsciously by someone else, but if we choose to be manipulated or design a system to do so ourselves, that's okay. Anyone who's eaten a meal standing over the kitchen sink—that is, everyone—knows that the same meal will be much more enjoyable if we sit ourselves at the dining room table and soak in the ambience.

We overemphasize money

Prices are just one of the many attributes that signal the value of things. They may be the only attribute that we can easily understand, but they're not the only attribute that matters. Consider using other criteria, even when they're hard to measure. We're all floating on that rough sea of uncertainty; don't let someone else's idea of value—that is, the price—be what you grab on to for salvation. A price is just a number, and while it can be a powerful part of a decision, it doesn't, and shouldn't, mean everything.

In general

When we don't have any specific idea about an item's value, we should do some research. Go online, investigate, ask around. With the massive amount of information available today—there's this thing called "the Internet"—we have no reason not to arm ourselves with knowledge. We don't need to spend a week researching the price of chewing gum, but we should probably dig around for a few hours, or at least a few minutes, before going to a car dealership.

WHAT'S IT GONNA TAKE TO GET YOU INTO THIS RESEARCH?

Car dealerships have a uniquely large asymmetry of information between the salespeople (who know a lot) and the rest of us (who know very little). Automobile salespeople frequently take advantage of that knowledge gap, and, as it happens, they are more likely to take advantage of certain consumers. Which ones? Women and minorities!

So, some people are more likely to benefit from doing online research before going to a car dealership than others. Who would gain more from arming themselves with information? The same groups: women and minorities.

Car dealerships are specifically tricky commercial settings, with many money traps and cultural biases, but the lesson here is general: Every time we face a situation where we know less than others and that gap can be used against us—which is the case in much of life and for people of all persuasions—we stand to gain a whole lot from studying up even a little bit.[1]

We want to be informed. Not just about our potential purchases, but about ourselves, our biases, and our money mistakes.

15

FREE ADVICE

Remember: Free is a price. It's a price that disproportionately grabs our attention.

The saying goes, "There's no such thing as free advice."

It's true: This chapter cost our publisher two pages.

16

CONTROL YOURSELF

Self-control is a matter that deserves special attention when we address how we think about money. Even if we manage to clear the many internal and external hurdles between us and a rational financial decision, a lack of self-control can trip us before we reach the finish line. We might be able to determine the correct value of our options, but our inability to control ourselves will end up nudging us to make the wrong choice.

Remember, our lack of self-control is due to discounting the future—because we are not emotionally attached to it—and to our willpower's failure to overcome the temptations of the present. So how can we increase self-control? By connecting to our future and resisting temptation. Easier said than done . . .

Back to the Future

We think of our future self as a somewhat separate person, so saving for the future can feel like giving money away to a stranger rather than giving it to ourselves.[1] One antidote is to reconnect to our future selves.

Hal Hershfield has been studying for a while all kinds of ways to overcome this flaw. In general, the findings amount to one powerful idea: Use simple tools to help us imagine our future self more vividly, specifically, and relatably.[2] It can be as simple as having an imaginary conversation with an older "us." Or we can write a letter to an elderly version of ourselves. We can also simply think about what our specific needs, desires, greatest joys, and toughest regrets will be when we're sixty-five, seventy, ninety-five, one hundred.

Talking with our future selves is one useful step toward shifting our thinking and building more willpower to resist the temptation of now. We don't need to have a sarcastic, negative discussion—"Oh no, young me didn't save. Now I live in a cardboard box!" It can and should be a positive and helpful one. Think about prepaying for a nice hotel. At check-in, we're told it's all paid for. We might turn toward younger us and say, "Hey, past me, you're a great guy for getting me this hotel! Awesome!" Now imagine that conversation when, instead of a prepaid hotel room, we leave ourselves $500,000 in a 401(k).

We can start with self-conversations, but we should also put in place other systems that help us become emotionally invested in our older selves. The more we can make the future defined, vivid, and detailed, the more relatable it becomes, and the more we'll care, connect, and act in our future selves' interests, too.

One way to become more invested is to change one of our most important decision-making environments: human resources. HR, the place where employees often make their savings decisions, should look like a doctor's office or retirement home. Or even better, like a doctor's office at a retirement home, decorated with bowls of hard candy, shuffleboard sticks, "Number 1 Grandma" mugs, and all kind of things that remind people of old age and long-term thinking. This is obviously more challenging for the growing millions of self-employed people in the world, but maybe we could dress up our kitchen table to look like an HR office when we are about to make retirement decisions.

One study found that people discounted the future less when it was described with a specific calendar date rather than as an amount of time. We are more likely to save for a retirement that happens on "October 18, 2037" than for one that happens "in twenty years." That simple change makes the future more vivid, concrete, real, and relatable.[3] That's an easy switch for HR professionals and investment advisors to make that can inspire us to save more.

We can also use technology to get people to connect to their future selves in a literal (and a little creepy) way. When we interact with computer-generated, old-age versions of ourselves, we save more.[4] We connect with the future old person. We experience empathy and emotion and we want to make this person's life easier. It doesn't matter if it's because of some sense of altruism toward others or raging self-interest, the result is the same: This person, this "future me," should be cared for.

This might seem like the plot of a sci-fi movie, but it's a powerful idea: Instead of *imagining* conversations with our older selves, we could actually have them; we could see and interact with a future us. Sure, we'd probably ask for winning lottery numbers and Super Bowl scores, but if that fails, we'd at least be more inclined to set aside more money for this person we now see in great detail. And look at us: We might also want to eat better and get some exercise. And moisturize, for goodness' sake; let's moisturize our skin.

Of course, most of us can't take virtual-reality tours of the future while we fill out benefits forms, so how can we democratize this idea of seeing our older self? Maybe our pay stubs or credit cards should have a picture of our face morphed to look older. Or, to tap into our aspirations and emotions about the future, we could use pictures of our older selves doing the wonderful things we could be doing in the perfect future—photos of hikes, vacations, playdates with grandkids, snapshots of our Olympic gold medals, presidential addresses, and space shuttle launches . . .

Tie Me to the Mast

When it comes to financial decision-making, we can try all sorts of things to make our present and future selves behave more in line with our long-term self-interest. One solution is to use binding self-control agreements, or what we call *ULYSSES CONTRACTS*.

We probably all remember the story of Ulysses and the Sirens. Ulysses knew that if the Sirens called to him, he would follow their voices to his and his sailors' doom, like so many sailors had done before him. He would not be able to control himself. But he wanted to hear the Sirens. (He'd been told that their latest album was "the bomb.") But, realizing that he couldn't resist their mythological beats, he asked his sailors to tie him to the mast of the boat. This way he could hear the call of the Sirens but could not act on his desire to follow them. In addition, he had his sailors put wax in their ears so that they couldn't even hear the Sirens or his pleas to be released, and wouldn't be tempted to sail to their doom. It worked. The ship survived.

A Ulysses contract is any arrangement by which we create barriers against future temptation. We give ourselves no choice; we eliminate free will. Unfortunately, Ulysses contracts rarely come with awesome music, but, on the other hand, they also rarely involve smashing our ship onto jagged rocks.

Common financial Ulysses contracts include things like preset limits on our credit cards or only using prepaid debit cards or even canceling all of those cards and only using cash. Another such pact has a decidedly non-Homeric name: "the 401(k)."

The Ulysses contract of a 401(k) is an irrational but remarkably effective strategy. The most rational approach to long-term saving is to wait until the end of each month and then look at our bills and expenses and, at that point, decide how much we can afford to save. But of course, if we follow this end-of-month strategy, we all know what will happen: We will never save, just like Rob Mansfield with his motorcycles and man caves. So what do we do? We pick an irrational strategy—precommitting to a type and quantity of savings, even though we don't know how much money we'll have or need each month. At least we're acknowledging our self-control failures and taking an action that will help us make the decisions we would like to make every month. The 401(k) (as well as other instruments like it) is certainly not an ideal strategy, but it's better than doing nothing. Importantly, this approach relies on a simple one-time decision that works for us in the long term: We only have to overcome temptation once, rather than twelve times a year. Overcoming one challenge is tough enough; overcoming twelve is even tougher. Reducing temptation is a good way to make better decisions, even if it isn't a good way to make reality television (the networks passed on Jeff's idea for "The Frugal Housewives & Rational Husbands of Overland Park").

Making retirement and savings contributions the automatic, default option, so that we must actively opt *out* of saving, is another wise approach. Not only does that eliminate the monthly, predictable problems of balancing saving for the future with the temptations and needs of the present, but we also eliminate even the one-time sign-up hurdle.

If we are automatically enrolled in a retirement savings plan, inertia and our tendency to be lazy work in our favor; they make us much more likely not to change anything and save for retirement in the first place. Later, they help us stay in the savings plan. Even though, logically, the decision about saving is just a decision about saving, and the two ways of approaching it should be identical—we're in or we're out—the effort required to sign up is enough to be a real barrier to saving. This concept of being automatically enrolled runs counter to traditional economic thinking—that we should and always can make informed, rational decisions—but runs right along the zigzaggedly too-human path of behavioral science.

When Rob was an employee in his twenties, his company made him actively choose to contribute, which he chose *not* to do. But what if he'd been automatically enrolled? He would probably not have taken an action to actively unenroll. The default option, combined with laziness and inertia, would have made a huge difference to his long-term savings.

These types of automatic savings plans—for retirement, college savings, health-care accounts, and the like—take the psychological traps that make automatic spending so prevalent (like the pain of paying and malleable mental accounting) and use them to our advantage. Automatic savings versus automatic spending: We *know* which one is a better choice, but when left to our own accord, we don't always choose it.

Ulysses contracts for savings really work. A study by Nava Ashraf, Dean Karlan, and Wesley Yin found that one group of participants who had their bank accounts restricted—that is, they chose to have money automatically deposited in a savings account—increased their savings by 81 percent within a year.[5]

Another study focused on automatically setting aside a portion of all *future salary increases*. That is, people automatically agreed to have a portion of their future raises set aside for savings. Their current earnings were not affected and they still got future raises, and when they did, these raises were just a little smaller. This practice also worked to increase savings. It's another great example of employing our psychological failures—in this case, the status quo bias and desire not to change anything—to overcome another—our lack of self-control.[6]

The process of earmarking is another way we can precommit ourselves to savings and encourage ourselves to stick to our plans. Earmarking—designating specific amounts of money for certain literal and mental accounts—can work to our advantage when it's a proactive, intentional decision (as opposed to the unintentional, knee-jerk reaction choices we discussed earlier, which cause problems). Earmarking can prevent us from using the money for all kinds of other purposes—especially ones that we did not plan to spend on from the get-go. We can earmark by using visual reminders on our pay stubs or setting money aside in separate bank accounts or—as we mentioned in the chapter about compartmentalization—we can put our weekly discretionary spending on a prepaid debit card.* Doing these things reminds us of the rules we've set up for ourselves and helps us keep ourselves "accountable." Pun intended.

We can manipulate ourselves further with emotional tricks like using nature's greatest tool: guilt. A study by Dilip Soman and Amar Cheema found that people were less likely to misuse earmarked money that was labeled with the names of their children than if their kids were left out of the process.[7] Yes, that's right: Envelopes full of cash that were labeled with the names of participants' children caused the parents to spend less and save more. How twisted, cruel, and, frankly, effective. Kids save the darnedest things.

We also might consider the ultimate financial Ulysses contract. Ulysses was tied to the mast. What if we took that binding and punishment further and created a discipline bank with a dominatrix as a logo? This bank would take every possible money decision out of our hands. Our employer would send our check to the discipline bank. The bank would pay our bills, and we'd get a weekly allowance. The money would be restricted. We couldn't do whatever we wanted with it, it would be set aside for specific usage, and the bank manager could change the rules as he or she saw fit. If we overdrew or otherwise violated our preset guidelines, we'd get punished, because we'd have been naughty, naughty. Heck, why not combine this with an earlier idea and have the bank logo be a picture of a dominatrix abusing a

*Would it be most beneficial to load money on to our weekly prepaid discretionary spending debit card on Monday or Friday? The answer is Monday. Why? Because if we do it Friday, we feel rich on the weekend, when we're more likely to spend without regard to our needs the following Wednesday or Thursday. If we load it on Monday, then we have a week of typically more set spending—transit to and from work, regular meals—and might plan and save more for the weekend splurge. The same logic can apply to the day of the week when we get our paycheck.

computer-generated older version of ourselves? We're sure that would get people to do . . . something . . . with their money.

Of course, we don't actually want this bank—whatever the logo—but we do wonder if we would enjoy living more without the need to worry all the time about managing our money. What if we were to farm out most of our decisions and responsibilities to a system, once, and then the system would manage money for us? Would we enjoy our lives a little more? We'd have less freedom, but also less worry. We think so, but we are not sure, so to test this out, send us all your money to hold on to and we'll see how it turns out. (We're kidding. Don't send us *all* your money.)

We should note that Ulysses contracts can be extremely useful tools to help us avoid temptation in almost every other part of our lives. Dan's undergrads tell him that during exam week they give their computers to one of their friends. They ask their friend to change their Facebook passwords so they can't log on again until exam period ends. Some of his female MBA students say that when they don't want a date to go too far, they wear ugly underwear. Perhaps we could even devise a literal Ulysses contract, where every time we give in to temptation we must read Homer's *Odyssey*, the epic poem about Ulysses. In the original Greek.

Treat Yourself

Another way to combat self-control problems is through *REWARD SUBSTITUTION*. Remember that one of our challenges is that we value a reward in the future—two marshmallows, a whole box of chocolates—much less than we value rewards in the present—even if the rewards in the present (one marshmallow, half a box of chocolates) are much, much smaller. What if we tried to bypass our inability to be motivated by future reward altogether and replaced it with another kind of present reward? Would that shift the balance toward greater self-control?

Dan had a particularly relevant experience in his complex medical life. As a teenager, Dan was hospitalized for severe burns. During that lengthy hospitalization, he contracted hepatitis C. Later, he was told of a Food and Drug Administration test to see if a new medication, interferon, could treat it. Dan joined the study, which unfortunately required him to take some unpleasant injections three times a week for a year and a half. Following every injection, he'd get extremely ill—shaking, fever, vomiting—for the whole night. If he completed the treatment, he'd reduce his chance of getting cirrhosis of the liver thirty years later . . . but he'd have to suffer tonight. It was an example of present sacrifice for future gains in a rather clear and extreme way.

Dan persevered and completed the treatment. He later found out that he was the only patient in the protocol to stick with the horrible medication regimen. He didn't manage to stick to the plan because he's some sort of superman or because he's better than us (this is where Jeff yells in the background, "He's not!"), but because he understood reward substitution.

Whenever he had to take this medication, he'd treat himself to a movie rental. He would get home, inject himself, and immediately start watching his highly anticipated movie, long before the bad side effects kicked in. He connected something unpleasant—the injection—with something pleasant—the movie. (From time to time he picked bad romantic comedies, which made him feel worse. We will publish *Dan's Top Movies for Overcoming Nausea* in the near future.)

Dan did not bother trying to connect to his future self. He didn't focus on the benefits of having a healthy liver. Those future benefits, while empirically important, couldn't compete with the present costs of the horrid side effects. Rather than teaching himself the importance of caring about his future, he changed his present environment. He gave himself a less important but far more immediate and tangible reason (the movies) to make a sacrifice today. Rather than focusing on the more important but less tangible reason (no more hep C) for the sacrifice, Dan focused on something much less important (a movie), but right now. That's reward substitution.

Maybe we could get people to spend more wisely and save more frequently if we offered reward substitutions for their rational behavior. Some states are doing just that by offering "lotteries" for people who put money into savings accounts.[8] Each deposit is greeted with a ticket that offers a small chance of winning an additional amount of money. These lottery-based savings plans work. Yet another example of reward substitution.

* * *

There are doubtless many other ways to combat issues of self-control, in many different situations. At a minimum, we must be aware that our lack of self-control always presents an obstacle to the success of even those brilliant financial decision-making systems we dissect in the pages to come.

17

IT'S US AGAINST THEM

A few pages ago, we discussed some tips for counteracting some of our many mental money miscues. We should recognize, however, that knowing how we *should* change our behaviors and actually changing them are two very different things. This is especially true with money, where we not only fight our own tendencies, but also fight a financial environment that actively tries to tempt us to make bad financial decisions. We live in a world where outside forces constantly want something from us—our money, our time, our attention—and that makes it hard to think rationally and act wisely.

For instance, we know that as long as mortgages were only described based on their interest rate, people could easily figure out which mortgage was a better deal, that is, 4 percent is less than 4.5 percent. (Even so, people don't spend much time trying to get cheaper mortgages. Many people don't understand that even a tiny decrease in the percentage—like from 3.5 to 3.25—adds up to big savings in the long run.)

But when mortgage brokers add a point system to their options—for example, we could pay an up-front amount of money, say $10,000, to reduce our interest payments by, say, 0.25 percent—our ability to compare the offers completely breaks down. Suddenly the calculation goes from one dimension (percentages) to two dimensions (up-front payment and percentages), and in this slightly more complex decision environment, we make more mistakes.

Now, you might say, "Oh, well, okay. Figuring out complex things is hard." True. But mortgage brokers are well aware of our difficulties calculating value when choices have multiple dimensions. So, presto! Suddenly mortgages are available with more and more options. These are presented as "consumer choices" and positioned as providing us the opportunity to make informed decisions . . . but, of course, more information and options means we can more easily make more mistakes. This is a system set up not to help us but rather to exacerbate our financial missteps.

So the struggle to improve our financial decision-making isn't just a struggle against our personal flaws; it's also against systems designed to exacerbate those flaws and take advantage of our shortcomings. Consequently, we must fight harder. We must individually adapt our thought processes to think more wisely about how we spend our money. And, as a society (assuming we want the people around us to make better money decisions), we must also design systems to be compatible with how we think about money so that our choices benefit us, and society, not those who might exploit and abuse our flawed thinking.

That's why the more we understand our flaws and limitations now, the better equipped we'll be to deal with them in the future. No one can predict the future: not about our investments, health, and jobs, nor about world events, celebrity presidents, and wine-drinking robots.*

What we do know is that the future will make our spending decisions even more challenging. From Bitcoin to Apple Pay, retinal scanners, Amazon preferences, and drone delivery, more and more modern systems are designed to make us spend more, more easily, and more

*Heck, thanks to a *Calvin and Hobbes* cartoon, Jeff thought he'd be playing saxophone for an all-girls cabaret in New Orleans by now.

often. We are in an environment that is ever more hostile to making thoughtful, well-reasoned, rational decisions. And because of these modern tools, it's only going to get more difficult for us to make choices that serve our long-term best interests.

The Temptation of Information

Now that we know that many commercial interests are after our time, our money, and our attention, we may think there's something we can do about it. After all, we believe ourselves to be reasonable and rational beings. So don't we just need the right information with which to make good decisions, and we will immediately make the right ones?

We eat too much? Just provide calorie information and all will be well. We don't save enough? Just start using a retirement calculator and watch our savings grow. Texting and driving? Just tell everyone how dangerous it is. Kids drop out of school? Doctors don't wash their hands before checking their patients? Let's just explain to the kids why they should stay in school and tell the doctors why they should wash their hands.

Sadly, life isn't that simple. Most of the problems we have in modern life are not due to lack of information, which explains why our repeated attempts to improve behavior by providing additional information often fail.

We're at an interesting inflection point in history, where technology can either work against us or for us. Currently, most financial technology is working against us, because most of it is designed to get us to spend more, sooner, rather than less, later. Technology is also designed to get us to think less about spending and to fail more frequently in the face of temptation. If we rely solely upon our instincts and the always-available technology, we are at the mercy of an overwhelming number of mechanisms that influence us to make the tempting short-term decision time after time.

For instance, the digital wallet is promoted as a pinnacle of modern consumer evolution. Free from cash, we can be flexible, save time, and focus less on managing our money while being provided with data to help us analyze our past spending. Sounds like a utopian era of technological bliss. Lines will be short, signatures will be quicker, access and enjoyment will be easier, faster, and frictionless. The hassle of payment will be eliminated and we'll enter a new, postmoney era of financial bliss.

Not so fast. More likely, these modern financial tools will further exacerbate our spending behaviors and we'll spend too much, too easily, too thoughtlessly, too fast, too often. This future looks bright if we're a bill collector or bankruptcy lawyer, but for most of us, that brightness comes from the flames burning a hole in our wallet.

It doesn't have to be like this.

More and more people recognize that the technology designed to make spending "easier" doesn't necessarily make it "better." People are starting to think not just about adjusting our behaviors, but changing our financial environment, our financial tools, and our financial default settings.

We can amplify our knowledge by designing systems, environments, and technologies that help us rather than tempt us. We can employ the very same behaviors and technologies that cause us harm to do us good. We can turn it all upside down on its head. We can use our quirks to our advantage.

How can we transform the financial environment? How can we create systems that are the opposite of Apple Pay and Android Pay—that is, instead of making spending more thoughtless, how do we help ourselves to think more clearly about it? Not just acting *after* we've finished something, like creating an accounting system that logs our expenses after they've been incurred, but creating a system that helps us before we make financial decisions in the first place? How? By rethinking what payment tools should look like for who we really are—people with limited time, attention, and cognitive capacity, and multiple quirks. By starting with an understanding of what we can and can't do well, we can design spending and saving instruments that could really help us.

We hope this book, the human flaws it exposes, and the handful of ways to use those flaws to our benefit will inspire all of us to take the next steps and develop such tools.

App-lied Psychologies

Consider the world of "apps." Unheard-of a decade ago, these are now today's hammers and screwdrivers. They are tools designed to entertain, educate, and enthrall. If apps can help us with physical fitness and mental well-being, why not financial fitness and fiscal well-being, too?

To keep track of opportunity costs, what if we developed an app that helped us do a bunch of comparisons and calculations all the time? It could automate the comparison: Thinking of $100 shoes? *Bing bong buzz!* Well, that's two movie tickets for you and your loved one, with popcorn and some wine after the film. Want to look good or feel good?

For managing both the good and bad aspects of mental accounting, what about apps that create categories and spending limits and then offer warnings when a limit for a category approaches?

To combat loss aversion, maybe we can develop an app that computes the expected value of our choices in a way that is independent from whether the choice is currently framed as a gain or loss. Want to sell your house? Maybe the app can help you set the right price and overcome your subjective attachment to it.

These are just a few starter ideas. The promising concept is that the same phones that we take with us everywhere could not just distract and tempt us, but could provide tools for better decisions in real time. Every coffee shop in Silicon Valley has a handful of unemployed coders waiting to help you develop more.

TOO MUCH OF A GOOD THING

There is a growing body of research that shows that too much information can hinder behavior change.[1] With apps monitoring sleep, heart rate, calories, exercise, steps, stairs, and breathing—not to mention spending and Internet use and other behaviors—we live in an age of personal quantification. We can instantly know how much of everything we're doing, have done, and should do. While it's great information to have, too much data can actually lessen the pleasure we get from even healthy activities, like exercise, sleep, diet, and savings. As data accumulates, and as we have to make an effort to measure, track, and think about it, the activities themselves can move from "lifestyle" to "work." As a consequence, our motivation to engage in these healthy activities drops. So, even if the data would help us understand what we *should* do, too much data defeats our desire to do anything about it.

As with all things—from wine and ice cream to technology and naps—moderation is key. Yes, even wine and ice cream should be consumed in moderation. (We didn't want to include that sentence, but our lawyers and doctors insisted.)

Scratch and Win

Since today's electronic wallets make us less aware of the pain of paying in an effort to increase *spending*, we could raise our spending *awareness*, which would increase the pain of paying, which would then reduce spending and increase *savings.*

We don't think about saving money very often. When we finally do think about it, our thoughts rarely lead us to save more. To test the extent that the design of digital wallets could influence behavior, Dan and his colleagues conducted a large-scale experiment with thousands of customers of a mobile money-saving system in Kenya. Some participants received two text messages every week: one at the start of the week to remind them to save and another one at the end of the week with a summary of their savings. Other participants got slightly different text reminders: It was framed like it came from their kid, asking them to save for "our future."

Four other groups were bribed (formally known as "financially incentivized") for saving. The first of these groups got a 10 percent bonus for the first 100 shillings that they saved. The second group got a 20 percent bonus for the first 100 shillings that they saved. The third and fourth groups got the same 10 percent and 20 percent bonuses for the first 100 shillings that they saved, but they got it together with loss aversion. (In these conditions, the researchers placed the full amount of the match—10 or 20 shillings—into their account at the beginning of the week. The participants were told that they would get the match based on how much they saved, and that the amount of the match that they did not save would be taken out of their account. Financially, this loss aversion approach was the same as the regular end-of-the-week match, but the idea was that experiencing money leaving their account would be painful and would get the participants to increase their savings.)

A final set of participants received those same text messages plus a golden-colored coin with the numbers 1–24 engraved on it, to indicate the 24 weeks that the plan lasted. These participants were asked to place the coin somewhere visible in their hut and scratch with a knife the number for that week to indicate if they saved or not.[2]

At the end of six months, the treatment that performed spectacularly better than every other was—drumroll please!—the coin. Every other treatment increased savings a bit, but those who received the coin saved about twice as much as those who only received text messages. You might think the winner would have been the 20 percent bonus or maybe the 20 percent bonus with loss aversion—and this is in fact what most people predict would be the most effective way to get people to save—but you'd be wrong.

How did a simple coin make such a substantial difference in behavior? Remember that participants received text message reminders to save. When you take into account the amount people saved on different days of the week, the results show that the coin did not get its advantage on the days when people got the reminders—it made its biggest impact on the other days. The gold coin made the act of saving salient by changing what people were thinking about as they were going about their day. From time to time they glimpsed the coin in their hut. Occasionally they touched it, talked about it, were aware of its presence. By being physically present, the coin brought the idea of saving, and with it the act of saving, into participants' daily lives. Not all the time, but now and then, and that was sufficient enough to get them to take action and make a difference.

This is a great example of how our thinking about money, about how our shortcomings, can be used to our benefit. We *should* react most strongly to the method that maximizes our money—a bonus for saving, which is free money—but we don't. We are more influenced by something that shapes our memory, attention, and thinking, such as the coin. Rather than lamenting that phenomenon as a financial personality disorder, we can design systems that provide us with the equivalent of a coin in many areas of life to motivate us to save more.

Showing Value

We can take this basic idea—that a physical representation of saving makes it more salient to the saver—and extend it into the community at large, by trying to adjust social values and gently pressuring people to save rather than to consume.

We often gauge the appropriate level of spending by watching what our peers and neighbors are doing—by eyeing their houses, cars, and vacations. These are things we can see. Savings, on the other hand, are not observable. Without prying or hiring a cadre of teenage Russian hackers, we don't know how much our colleagues put into their 401(k), only, in general, how much they put into new clothes, kitchen renovations, and cars. Due to our awareness, we experience social pressure to "keep up with the Joneses" on spending, but not on the invisible savings.

Consider other cultures. In some places in Africa, people save by buying more goats. If we are doing well, we have more goats on our property, and everybody knows how many goats we have. There are other places where people save by buying bricks, so they pile bricks outside their hut until they have gathered enough to build another room. In this case, too, other people know how many bricks everyone has.

When it comes to savings, there is nothing similar in our modern digital culture. When we put money into a college savings account or a 401(k), we don't get trumpeting fanfare or a brighter set of holiday lights. When we buy our child a gift, they know we did it and can be thankful for it. Not so when we put that money into their 529 college savings account.

So how do we make these "invisible things" visible, not just so our good behaviors are appreciated, but also to start a conversation about saving among families and communities? So that we can gain the support of others in making financial sacrifices for the future that are, too often, done in near silence and secrecy?

When we perform our civic duty at the ballot box, we get a sticker that says "I Voted." When democracy recently came to countries like Iraq and Afghanistan, citizens there proudly held up their purple ink-stained fingers as a sign of participation. Could there be something similar for doing the duty of saving? Something to show what types of accounts we have opened for saving for ourselves and for our kids?

Could we get stickers when we have saved more than 15 percent of our income? Small trophies? Large statues? Scarlet dollar signs on our lapels and our homes? It would be tacky to have one of those big thermometers outside our home marking each savings milestone, but there's no question that if we did, we'd all save more. Until we make such meters culturally acceptable, maybe we could start celebrating when we pay off our mortgages or car payments? Instead of a Sweet 16 party, it would be a Sweet, Now I Can Afford to Send My 16-Year-Old to College party.

These ideas may not be practical, but the principle of making invisible savings visible is something we should build on. We can start by encouraging conversations about what's reasonable to save so that we compete not just for bigger cars, but also for bigger savings.

SEE HOW GOOD I AM?

The benefits of displaying our wise decisions and altruistic choices aren't confined to the world of finance. Celebrations of good behavior could be useful in other parts of our lives as well.

Consider global warming. Outside of recycling and the occasional yelling at the news, few of us make regular personal sacrifices for the benefit of the earth's future. What if we were to use reward substitution to display the value of such decisions? Could we, essentially, get people to do the right thing for the wrong reasons? Well . . . yes. We could and we do.

Think about the Toyota Prius and the Tesla. These cars allow their drivers to communicate to the rest of us what generous, wonderful, caring, better-than-you people they are. Prius and Tesla drivers can smile and look at *themselves* and think, "I am a fine human being." They can also show the world that they've made this decision and they believe that other people look at them and their cars and say, "Oh, what a fine human being must be driving this ecological masterpiece!" The direct reward of fighting global warming might not be enough for everyone, but if it is combined with this ego stoking, well, maybe it will get more people to care about slowing the rising tides for a day or two.

I Believe the Children Have a Future

Research shows that when parents open a college savings account for their kids, those kids perform better throughout their lives. Some states are combining this finding with the equally important finding that if poor people are given some assets, they start saving and have better financial futures. The endowment effect, loss aversion, mental accounting, and anchoring are some of the mechanisms that contribute to these positive outcomes.

Child development accounts (aka CDAs) are savings or investment accounts designed for long-term developmental purposes. These programs provide new parents with an automatic college savings account, an initial deposit of $500 or $1,000, a savings match, account statements, regular information about college, and reminders about saving for college.

Why do these programs work? For many of the same reasons that the gold coin worked. In addition to helping families save money, CDAs work on our psychology. They remind parents and children that college is an attainable, perhaps even anticipated, part of life and that saving for it is important. Account statements let families know the state of their asset growth. In addition, children who know they have the ability and tools to attend college become more hopeful about doing so, more focused and more future-oriented toward these goals. And finally, these kids and their parents are more likely to develop expectations and an identity around the concept of attending college themselves.[3]

CDAs are another example of an intentionally designed financial environment that values saving and the mindset that goes along with it. CDAs remind people about savings, provide a sense of ownership, and help people overcome worries about giving up some money now by highlighting the long-term value of their goals. All of this ever so slightly tilts the psychology of money into our favor.

Check This Out

Most people live with a fixed amount of income—salary, benefits, etc.—and a certain level of fixed expenses—housing, transportation, insurance, and so forth. The rest is what we call "discretionary." We should feel comfortable spending part of this discretionary amount, but we should also avoid touching some of it, and instead recategorize it as savings, delayed spending, or rainy-day funds.

The method we use to determine what portion of our discretionary money goes into which category—"easy to spend" or "off-limits"—can be used to our advantage. Currently, the simplest way we measure our discretionary money is by how much money we have in our checking account, that is, our checking balance. If we have less in our checking—or we *feel* that we have less in our checking—it constrains our spending behavior. If we feel we have a higher balance, we go ahead and spend more.

There are several ways to use this checking balance rule to our advantage, to use it to trick ourselves into saving. For example, we can move a little bit of money out of our checking and into a savings account. That way, our checking account will be artificially too low and it will get us to think that we're poorer than we really are. We could produce a similar outcome by asking our employer to direct-deposit some of our salary into separate accounts, to help us "forget" about these savings. With approaches like these, we would still use the balance in our checking as an indication of how much we should spend, but we'd find ourselves buying one or two fewer dinners or special treats and we'd reduce our overall spending.

Essentially, we can spend less by hiding money from ourselves. Yes, if we stop to consider it, we know we're hiding it and where. But we can take advantage of our cognitive laziness and the fact that we don't regularly think about how much money is in our other accounts—and we think about it even less if it's automatic deposit and we don't move the money ourselves every time. So, tricking ourselves is an easy and useful strategy. It wouldn't be a permanent deceit, but it would surely prevent some irrational purchases.

More Power to Ya

There are many more tricks we can use to save money. For instance, in the United Kingdom, some people have the option of putting coins into a meter whenever they want to heat their homes, thereby harnessing the mental power of the pain of paying to reduce their power bills. Rather than someone reading the meter monthly and then issuing you a bill, and then a little bit later you pay for it . . . these Brits frequently feel the psychological pain of paying for a little more warmth. Then they can decide to just put on a sweater.

Moving from those who pinch pennies to those who have enough pennies to forget about some of them . . . experts at Fidelity Investments recently learned that the investors whose portfolios performed the best were those who had completely forgotten that they had investment portfolios at all.[4] That is, the investors who simply left their investments alone—without trying to trade or manage, without getting trapped by tendencies to herd, overemphasize price, be loss averse, overvalue what they own, and fall victim to expectations—did the best. By making a "smart investment" choice, then leaving it alone, they minimized their money mistakes. We can do that, too. We can also dream that somewhere there's a large investment account we've forgotten about. . . .

It should be noted that some successful investors left their investments alone because they died. That suggests that "playing dead" isn't just a good way to avoid bear attacks; it's also a sound investment strategy. (There's probably a "bear market" lesson in here, too, but it's getting late in the book, so let's move on.)

The Illusion of Wealth

We react differently to "Oh, this coffee is $4 a day" than to "Oh, this coffee is $1,460 per year." How we describe the time frame in which an amount of money is spent—in hours, weeks, months, or years—has a huge impact on how we think about the value and wisdom of our spending decisions.

In one set of experiments, when we gave people a salary of $70,000 but framed it as hourly earnings of $35 an hour, they saved less than when we defined it as a yearly sum of $70,000. When our salary is presented as a yearly amount, we take a more long-term view. Consequently, we save more for retirement. Of course, in the United States most low-income jobs are paid by the hour, which typically worsens the problem of not saving for the long term.

This phenomenon by which a lump sum of $100,000 at retirement seems larger than its equivalent of approximately $500 per month for life is called the "illusion of wealth."[5] And while the "illusion of wealth" can be seen as a flaw in our thinking, it can also be something that we can use to design saving systems to our advantage. In the case of retirement savings, stating retirement income in monthly terms should therefore make us feel that we are saving less than we need, and make us think that we should increase the amount and save more. Similarly, we could put projected monthly income at our expected time of retirement before any other information on our 401(k) statements, making it salient that the need for savings is still high. Some retirement plans have already taken steps in these directions, with positive outcomes.[6]

Once we better understand such quirks in the ways we think about numbers, we can figure out how to use them to our long-term advantage and change our savings behavior and choices. It seems that using the right time frame is an important factor. To persuade people to take money out of their salary, we should frame their earnings yearly. To persuade them that they'll need more savings in the future, we should frame their spending monthly. That dominatrix we mentioned earlier might help, too.

In addition to these number-framing devices, there are other useful ways to handle our year-to-year income that can increase our happiness and curtail poor spending decisions. When we have a regular income—say, $5,000 per month—we tend to scale up our life expenses to fit in range of this $5,000. What if, on top of that, we gave ourselves a bonus? How would we use this money?

At some point, Dan asked his students to imagine that they worked for him and could either get a raise of $1,000 per month or $12,000 as an end-of-year bonus. Virtually everyone agreed that a monthly increase would be more rational. For one thing, they would get the money sooner. Everyone also said they'd use the money differently if it were a monthly increase as opposed to an end-of-year bonus. If they got it every month, it would be part of the regular flow of money, and they'd use it for mundane things such as bills and monthly expenses. But if they got it at the end of the year, it would not be part of the mental account that comes with salary. Thus they would feel more free to spend it on special purchases that would provide more happiness than simply paying the bills. Now, hopefully, not all of the $12,000 would be spent this way, but some of it would be used more freely.

So, if the choice is between a salary of $6,000 per month versus a salary of $5,000 per month plus a $12,000 end-of-year bonus, what would happen to the quality of life? The $6,000-per-month "person" would probably increase his or her quality of life with a slightly better car, apartment, meals, but would not be able to do something big for themselves. Whereas the person with the bonus would be able to do special things like buy a motorcycle, pay for a vacation, or start a savings account.

This may seem to contradict what we've just said about lump sums and savings, but 1) that was savings, this is spending, 2) we are human, and 3) no one ever blamed human behavior for being consistent.

People use the phrase "pay yourself first" for savings, and we should. But if we have relatively stable income, one useful way of getting more joy out of it is to shave off some of that regular income, adjust our expenses to the lower standard spending amount, and use that shaved-off savings—the shavings, if you will—to give ourselves a bonus. Then we could use some of that bonus on something we'd truly, truly enjoy. Yes, we should pay our future selves first, but we can shave off a little for our present selves, too.

18

STOP AND THINK

The last few chapters provided just a few examples of designing environments to turn our mental shortcomings into tools that work in the service of our financial success.

We could go on and on, picking out experiments and efforts from around the globe, but the point is this: Work has started in an attempt to use our human quirks—as revealed by financial psychology and behavioral economics—to improve the outcomes of our flawed thinking, as opposed to just taking advantage of it. Given what we see out in the real world, however, it is clear that much more needs to be done.

It would be fantastic if we were able to design more systems like these to improve our financial environments, reduce the impact of our mental money mistakes, and weaken the outside forces that lead us astray.

But the truth is, these forces aren't our only, or our biggest, enemies: We are. If we didn't make poor value judgments in the first place, we wouldn't be able to be exploited to the degree we are now. We need to understand and accept our flaws and shortcomings. Don't believe everything you think. Stop being stubborn. Don't assume you're too smart to fall for these kinds of tricks or that they only work on other people.

A wise man knows himself to be a fool, but a foolish man opens his wallet and removes all doubt.

Recognizing that we respond to irrelevant value cues gives us the opportunity to learn, grow, and improve as financial individuals, and to have more money to celebrate that growth (hopefully delaying a bit the celebration).

The amazing cartoonist Sam Gross drew a panel where two men stand in front of a giant billboard with the words STOP AND THINK. One man turns to the other and says, "It sort of makes you stop and think, doesn't it?"

We need that type of road sign to interrupt us on our financial journeys, to wake us up from our financial sleepwalking. And we need that sign to appear pretty often, just to provide a moment, a pause, some additional friction, something to take us off automatic, keep us present, and help us consider what we're doing.

If we sit on the couch with one large bag of popcorn or crackers, we're going to eat it all mindlessly. If, however, we're given the same overall amount but divided into four smaller bags, we pause in the moment when we have to switch to a new bag. This small action gives us an opportunity to reflect and decide whether we want to eat more or not. As it turns out, with the breaks afforded by multiple bags, we eat fewer snacks than when we're given just the large bag.

Translating that snacking tendency into the world of finance, if we get all of our money for a given period in one big envelope, we tend to just spend it all, as mindlessly as eating on the couch. But if the same amount is split into multiple envelopes, we halt our spending at the end of each one. Furthermore, as we noted before, if we take these envelopes and put the names of our children on them, we're even less likely to keep spending.[1]

The reason we adjust our snacking or spending when it's time to open a new bag or envelope is that the act of opening that new container forces us to pause and think about what we are doing. That creates a decision point, during which we evaluate, ever so slightly, our actions and reconsider our next steps.

Throughout this book, we've tried to show that we face many decisions in our financial lives. We often don't pause to think about these decisions, and we often don't even realize that these are, in fact, decisions to face and make at all. Yet we make a lot of financial decisions, and in many of those we receive numerous irrelevant value cues, to which we respond time and again. These are the things about which we need to become more aware. Then we might, from time to time, stop and think—and maybe make better decisions.

Life is full of decisions. Big decisions, small decisions, and repeated decisions. The big decisions—like buying a house or getting married or choosing a college—are places where it makes sense to stop and think as much as possible about value and spending. Most of us do that. Not enough, but at least we do it some.

The small decisions—like splurging on a treat during the county fair or an extra dish at your anniversary dinner—aren't generally worth the time and effort to worry about value cues. Yes, it would be nice to think about these, but adopting this kind of thinking about every small decision all the time would drive anyone mad.

Then there are the repeated decisions, which are essentially small decisions we make over and over again. They're habits, like buying coffee, shopping at the supermarket, going out to eat, or buying flowers for your loved one every week. Each purchase is individually small, but we make them a lot, so they have a large cumulative impact. We still probably shouldn't dwell on each of these repeated purchases every time, but now and then, maybe at the end of a semester or season or book, we can stop and think about them. (Obviously, we were just joking about buying flowers—we have yet to meet someone who is spending enough on showing love to their significant other.)

* * *

So, we're not saying we should question every financial decision, always, in every way possible. That would be economically sound, but psychologically overwhelming, daunting, and unwise. We don't want to become frightened, stingy, or constantly worried. So don't question everything. Life is meant to be enjoyed. But pick your spots and question those things that are most likely to cause long-term harm.

Every so often, consider how much pleasure, how much value, we may truly get out of a purchase. Think about what else we could spend that money on and why we're making this choice. If we recognize what we're doing and why, over time, slowly but surely, we'll get the ability to change our decision-making for the better.

Money is a difficult and abstract concept. It is hard to deal with and hard to think about. But that doesn't mean we're helpless. So long as we understand incentives and tools and our own psychology, we can fight back. If we're willing to dig deeper into human psychology, we might improve our behavior, our lives, and our freedom from financial confusion and stress.

Money Is Important and Foolish . . . and So Are We

Jeff was once paid to write a campaign speech for someone running for the powerful political position of fifth-grade student council. (She won; otherwise he wouldn't be sharing this story.) He spent most of his time on the job reassuring the parents—successful hedge fund managers—that they were good people, even though he actually thought that their wealth and relationship with money had distorted their values and their relationship with their child as well. So why did he fib to them? Why did he even take the job? For the money, of course. (He likes to say it was "for the story," but really, it was mainly for the money.)

Money makes everyone do crazy things. And if we've learned anything from broke lottery winners and bankrupt professional athletes, even having lots of money doesn't make it easier to think about. Sometimes, quite the opposite.

So what should we do? We could try to abandon the modern economy and find ways around money. We could go to a basket-weaving commune or start a money-free, barter-based community where every meal costs a single Albanian three-toed blork. But then we'd miss out on theater and art and travel and wine. Money has allowed us to develop the vast, intricate, and amazing modern society that we all share, that makes life worth living and money worth earning.

So, let's find a peaceful coexistence with money. There is a growing movement of billionaires giving away their fortunes, recognizing the value of charity and the negative effect of extreme wealth. There is also a growing body of literature on how to get more pleasure, meaning, and fulfillment from our spending (led by our friends Mike Norton and Elizabeth Dunn and their book, *Happy Money*). You probably have some good ideas yourself. Share them, develop them, explore their possibilities. Let's keep thinking about money and how we can find a harmonious coexistence with this tricky yet vital invention.

It's also essential that we all start talking to our friends about money. It is not easy to talk about what we do with money, how much we save, how much we spend, and the many money mistakes we make. But it is important for us to help each other deal with money and the complex decisions about it that we face.

In the end, money really isn't the only thing that matters. But it does matter, to all of us, a lot. We spend an extraordinary amount of time thinking about it—and often thinking about it incorrectly.

We could continue to let the price setters, salespeople, and commercial interests take advantage of our psychology and behavior and tendencies and foolishness. We could wait for societal or governmental interests to put programs in place to protect us from our own foolishness. Or we could become more aware of our limitations, design personal systems to correct ourselves, and take control of our financial decisions so that our precious, finite, and immeasurably valuable lives can grow richer every day.

It's up to us. We raise our dirty coffee mugs of delicious wine in a toast to a better tomorrow.

Cheers,

Dan and Jeff

THANKS

Dan and Jeff would like to extend their heartfelt gratitude to money. Thank you for being so complex. Thank you for all the ways you make it difficult to think about you. Thank you for allowing the financial world to become extra complex.

Thank you for credit cards, mortgages, hidden fees, mobile banking, casinos, car dealerships, financial advisers, Amazon.com, real estate listings, the fine print, and apples and oranges.

Without you, life would be much simpler, but there would be no need for this book.

This book would be full of mere speculation if not for the brilliant work of the researchers, professors, and authors cited within these pages.

It would also be a jumble of nonsense words without the immense talents of Elaine Grant, Matt Trower, and Ingrid Paulin.

And it would just be a corrupt file on our hard drives without the love and support of Jim Levine and the insight and passion of Matt Harper.

We thank you all.

Jeff would also like to thank his parents, because that's what ungrateful kids do; his siblings, for being trailblazers in the field of ungratefulness; his wife, Anne, for her patience, inspiration, and love; his kids, Scott and Sarah, for having the best laughs in the world; and, of course, Dan Ariely, for using his Israeli accent—which somehow hasn't faded after decades in America—to pierce the noise at a restaurant in North Carolina and ask, "So, maybe we should write something about money?"

Dan Ariely also loves his family, but he prefers to leave the details to your imagination.

NOTES

Introduction

1. Kathleen D. Vohs (University of Minnesota), Nicole L. Mead (Florida State University), and Miranda R. Goode (University of British Columbia), "The Psychological Consequences of Money," *Science* 314, no. 5802 (2006): 1154–1156.

2. Institute for Divorce Financial Analysts, "Survey: Certified Divorce Financial Analyst® (CDFA®) Professionals Reveal the Leading Causes of Divorce," 2013, https://www.institutedfa.com/Leading-Causes-Divorce/.

3. Dennis Thompson, "The Biggest Cause of Stress in America Today," CBS News, 2015, http://www.cbsnews.com/news/the-biggest-cause-of-stress-in-america-today/.

4. Anandi Mani (University of Warwick), Sendhil Mullainathan (Harvard), Eldar Shafir (Princeton), and Jiaying Zhao (University of British Columbia), "Poverty Impedes Cognitive Function," *Science* 341, no. 6149 (2013): 976–980.

5. Paul K. Piff (UC Berkeley), Daniel M. Stancato (Berkeley), Stéphane Côté (Rotman School of Management), Rodolfo Mendoza-Denton (Berkeley), and Dacher Keltner (Berkeley), "Higher Social Class Predicts Increased Unethical Behavior," *Proceedings of the National Academy of Sciences* 109 (2012).

6. Maryam Kouchaki (Harvard), Kristin Smith-Crowe (University of Utah), Arthur P. Brief (University of Utah), and Carlos Sousa (University of Utah), "Seeing Green: Mere Exposure to Money Triggers a Business Decision Frame and Unethical Outcomes," *Organizational Behavior and Human Decision Processes* 121, no. 1 (2013): 53–61.

Chapter 2: Opportunity Knocks

1. Shane Frederick (Yale), Nathan Novemsky (Yale), Jing Wang (Singapore Management University), Ravi Dhar (Yale), and Stephen Nowlis (Arizona State University), "Opportunity Cost Neglect," *Journal of Consumer Research* 36, no. 4 (2009): 553–561.

Chapter 3: A Value Proposition

1. Adam Gopnik, Talk of Town, "Art and Money," *New Yorker*, June 1, 2015.

2. Jose Paglieri, "How an Artist Can Steal and Sell Your Instagram Photos," CNN, May 28, 2015, http://money.cnn.com/2015/05/28/technology/do-i-own-my-instagram-photos/.

Chapter 4: We Forget That Everything Is Relative

1. Brad Tuttle, "JC Penney Reintroduces Fake Prices (and Lots of Coupons Too, Of Course)," *Time*, May 2, 2013, http://business.time.com/2013/05/02/jc-penney-reintroduces-fake-prices-and-lots-of-coupons-too-of-course/.

2. Brian Wansink, *Mindless Eating: Why We Eat More Than We Think* (New York: Bantam, 2010).

3. Aylin Aydinli (Vrije Universiteit, Amsterdam), Marco Bertini (Escola Superior d'Administració i Direcció d'Empreses [ESADE]), and Anja Lambrecht (London Business School), "Price Promotion for Emotional Impact," *Journal of Marketing* 78, no. 4 (2014).

Chapter 5: We Compartmentalize

1. Gary Belsky and Thomas Gilovich, *Why Smart People Make Big Money Mistakes and How to Correct Them: Lessons from the New Science of Behavioral Economics* (New York: Simon & Schuster, 2000).

2. Jonathan Levav (Columbia) and A. Peter McGraw (University of Colorado), "Emotional Accounting: How Feelings About Money Influence Consumer Choice," *Journal of Marketing Research* 46, no. 1 (2009): 66–80.

3. Ibid.

4. Amar Cheema (Washington University, St. Louis) and Dilip Soman (University of Toronto), "Malleable Mental Accounting: The Effect of Flexibility on the Justification of Attractive Spending and Consumption Decisions," *Journal of Consumer Psychology* 16, no. 1 (2006): 33–44.

5. Ibid.

6. Eldar Shafir (Princeton) and Richard H. Thaler (University of Chicago), "Invest Now, Drink Later, Spend Never: On the Mental Accounting of Delayed Consumption," *Journal of Economic Psychology* 27, no. 5 (2006): 694–712.

Chapter 6: We Avoid Pain

1. Donald A. Redelmeier (University of Toronto), Joel Katz (University of Toronto), and Daniel Kahneman (Princeton), "Memories of Colonoscopy: A Randomized Trial," *Pain* 104, nos. 1–2 (2003): 187–194.

2. Drazen Prelec (MIT) and George Loewenstein (Carnegie Mellon University), "The Red and the Black: Mental Accounting of Savings and Debt," *Marketing Science* 17, no. 1 (1998): 4–28.

3. Nina Mazar (University of Toronto), Hilke Plassman (Institut Européen d'Administration des Affaires [INSEAD]), Nicole Robitaille (Queen's University), and Axel Lindner (Hertie Institute for Clinical Brain Research), "Pain of Paying? A Metaphor Gone Literal: Evidence from Neural and Behavioral Science," INSEAD Working Paper No. 2017/06/MKT, 2016.

4. Dan Ariely (MIT) and Jose Silva (Haas School of Business, UC Berkeley), "Payment Method Design: Psychological and Economic Aspects of Payments" (Working Paper 196, 2002).

5. Prelec and Loewenstein, "The Red and the Black."

6. For a review: Dilip Soman (University of Toronto), George Ainslie (Temple University), Shane Frederick (MIT), Xiuping L. (University of Toronto), John Lynch

(Duke University), Page Moreau (University of Colorado), George Zauberman (UNC Chapel Hill), et al., "The Psychology of Intertemporal Discounting: Why Are Distant Events Valued Differently from Proximal Ones?" *Marketing Letters* 16, nos. 3–4 (2005): 347–360.

7. Elizabeth Dunn (University of British Columbia) and Michael Norton (Harvard Business School), *Happy Money: The Science of Happier Spending* (New York: Simon & Schuster, 2014): 95.

8. Drazen Prelec (MIT) and Duncan Simester (MIT), "Always Leave Home Without It: A Further Investigation of the Credit-Card Effect on Willingness to Pay," *Marketing Letters* 12, no. 1 (2001): 5–12.

9. Richard A. Feinberg (Purdue), "Credit Cards as Spending Facilitating Stimuli: A Conditioning Interpretation," *Journal of Consumer Research* 12 (1986): 356–384.

10. Promotesh Chatterjee (University of Kansas) and Randall L. Rose (University of South Carolina), "Do Payment Mechanisms Change the Way Consumers Perceive Products?" *Journal of Consumer Research* 38, no. 6 (2012): 1129–1139.

11. Uri Gneezy (UC San Diego), Ernan Haruvy (UT Dallas), and Hadas Yafe (Israel Institute of Technology), "The Inefficiency of Splitting the Bill," *Economic Journal* 114, no. 495 (2004): 265–280.

Chapter 7: We Trust Ourselves

1. Gregory B. Northcraft (University of Arizona) and Margaret A. Neale (University of Arizona), "Experts, Amateurs, and Real Estate: An Anchoring-and-Adjustment Perspective on Property Pricing Decisions," *Organizational Behavior and Human Decision Processes* 39, no. 1 (1987): 84–97.

2. Amos Tversky (Hebrew University) and Daniel Kahneman (Hebrew University), "Judgment under Uncertainty: Heuristics and Biases," *Science* 185 (1974): 1124–1131.

3. Joseph P. Simmons (Yale), Robyn A. LeBoeuf (University of Florida), and Leif D. Nelson, (UC Berkeley), "The Effect of Accuracy Motivation on Anchoring and Adjustment: Do People Adjust from Provided Anchors?" *Journal of Personality and Social Psychology* 99, no. 6 (2010): 917–932.

4. William Poundstone, *Priceless: The Myth of Fair Value (and How to Take Advantage of It)* (New York: Hill & Wang, 2006).

5. Simmons, LeBoeuf, and Nelson, "The Effect of Accuracy Motivation on Anchoring and Adjustment."

6. Dan Ariely (Duke University), *Predictably Irrational* (New York: HarperCollins, 2008).

Chapter 8: We Overvalue What We Have

1. Daniel Kahneman (Princeton), Jack L. Knetsch (Simon Fraser University), and Richard H. Thaler (University of Chicago), "The Endowment Effect: Evidence of Losses Valued More than Gains," *Handbook of Experimental Economics Results* (2008).

2. Michael I. Norton (Harvard Business School), Daniel Mochon (University of California, San Diego), and Dan Ariely (Duke University), "The IKEA Effect: When Labor Leads to Love," *Journal of Consumer Psychology* 22, no. 3 (2012): 453–460.

3. Ziv Carmon (INSEAD) and Dan Ariely (MIT), "Focusing on the Forgone: How Value Can Appear So Different to Buyers and Sellers," *Journal of Consumer Research* 27, no. 3 (2000): 360–370.

4. Daniel Kahneman (UC Berkeley), Jack L. Knetsch (Simon Fraser University), and Richard Thaler (Cornell), "Experimental Tests of the Endowment Effect and the Coarse Theorem," *Journal of Political Economy* 98 (1990): 1325–1348.

5. James R. Wolf (Illinois State University), Hal R. Arkes (Ohio State University), and Waleed A. Muhanna (Ohio State University), "The Power of Touch: An Examination of the Effect of Duration of Physical Contact on the Valuation of Objects," *Judgment and Decision Making* 3, no. 6 (2008): 476–482.

6. Daniel Kahneman (University of British Columbia) and Amos Tversky (Stanford), "Prospect Theory: An Analysis of Decision under Risk," *Econometrica: Journal of Econometric Society* 47, no. 2 (1979): 263–291.

7. Belsky and Gilovich, *Why Smart People Make Big Money Mistakes.*

8. Dawn K. Wilson (Vanderbilt), Robert M. Kaplan (UC San Diego), and Lawrence J. Schneiderman (UC San Diego), "Framing of Decisions and Selection of Alternatives in Health Care," *Social Behaviour* 2 (1987): 51–59.

9. Shlomo Benartzi (UCLA) and Richard H. Thaler (University of Chicago), "Risk Aversion or Myopia? Choices in Repeated Gambles and Retirement Investments," *Management Science* 45, no. 3 (1999): 364–381.

10. Belsky and Gilovich, *Why Smart People Make Big Money Mistakes.*

11. Hal R. Arkes (Ohio University) and Catherine Blumer (Ohio University), "The Psychology of Sunk Cost," *Organizational Behavior and Human Decision Processes* 35, no. 1 (1985): 124–140.

Chapter 9: We Worry About Fairness and Effort

1. Alan G. Sanfey (Princeton), James K. Rilling (Princeton), Jessica A. Aronson (Princeton), Leigh E. Nystrom (Princeton), and Jonathan D. Cohen (Princeton), "The Neural Basis of Economic Decision Making in the Ultimatum Game," *Science* 300 (2003): 1755–1758.

2. Daniel Kahneman (UC Berkeley), Jack L. Knetsch (Simon Fraser University), and Richard H. Thaler (Cornell), "Fairness as a Constraint on Profit Seeking: Entitlements in the Market," *American Economic Review* 76, no. 4 (1986): 728–741.

3. Annie Lowrey, "Fare Game," *New York Times Magazine*, Jan. 10, 2014.

4. On Amir (UC San Diego), Dan Ariely (Duke), Ziv Carmon (INSEAD), "The Dissociation Between Monetary Assessment and Predicted Utility," *Marketing Science* 27, no. 6 (2008): 1055–1064.

5. Jan Hoffman, "What Would You Pay for This Meal?" *New York Times*, Aug. 17, 2015.

6. Ryan W. Buell (Harvard Business School) and Michael I. Norton (Harvard Business School), "The Labor Illusion: How Operational Transparency Increases Perceived Value," *Management Science* 57, no. 9 (2011): 1564–1579.

Chapter 10: We Believe in the Magic of Language and Rituals

1. John T. Gourville (Harvard) and Dilip Soman (University of Colorado, Boulder), "Payment Depreciation: The Behavioral Effects of Temporally Separating Payments From Consumption," *Journal of Consumer Research* 25, no. 2 (1998): 160–174.

2. Nicholas Epley (University of Chicago), Dennis Mak (Harvard), and Lorraine Chen Idson (Harvard Business School), "Bonus or Rebate? The Impact of Income Framing on Spending and Saving," *Journal of Behavioral Decision Making* 19, no. 3 (2006): 213–227.

3. John Lanchester, *How to Speak Money: What the Money People Say—and What It Really Means* (New York: Norton, 2014).

4. Kathleen D. Vohs (University of Minnesota), Yajin Wang (University of Minnesota), Francesca Gino (Harvard Business School), and Michael I. Norton (Harvard Business School), "Rituals Enhance Consumption," *Psychological Science* 24, no. 9 (2013): 1714–1721.

Chapter 11: We Overvalue Expectations

1. Elizabeth Dunn (University of British Columbia) and Michael Norton (Harvard Business School), *Happy Money: The Science of Happier Spending* (New York: Simon & Schuster, 2014).

2. Michael I. Norton (MIT) and George R. Goethals, "Spin (and Pitch) Doctors: Campaign Strategies in Televised Political Debates," *Political Behavior* 26 (2004): 227.

3. Margaret Shin (Harvard), Todd Pittinsky (Harvard), and Nalini Ambady (Harvard), "Stereotype Susceptibility Salience and Shifts in Quantitative Performance," *Psychological Science* 10, no. 1 (1999): 80–83.

4. Ibid.

5. Robert Rosenthal (UC Riverside) and Leonore Jacobson (South San Francisco Unified School District), *Pygmalion in the Classroom: Teacher Expectation and Pupils' Intellectual Development* (New York: Holt, Rinehart & Winston, 1968).

6. James C. Makens (Michigan State University), "The Pluses and Minuses of Branding Agricultural Products," *Journal of Marketing* 28, no. 4 (1964): 10–16.

7. Ralph I. Allison (National Distillers Products Company) and Kenneth P. Uhl (State University of Iowa), "Influence of Beer Brand Identification on Taste Perception," *Journal of Marketing Research* 1 (1964): 36–39.

8. Samuel M. McClure (Princeton), Jian Li (Princeton), Damon Tomlin (Princeton), Kim S. Cypert (Princeton), Latané M. Montague (Princeton), and P. Read

Montague (Princeton), "Neural Correlates of Behavioral Preference for Culturally Familiar Drinks," *Neuron* 44 (2004): 379–387.

9. Moti Amar (Onno College), Ziv Carmon (INSEAD), and Dan Ariely (Duke), "See Better If Your Sunglasses Are Labeled Ray-Ban: Branding Can Influence Objective Performance" (working paper).

10. Belsky and Gilovich, *Why Smart People Make Big Money Mistakes*, 137.

11. Baba Shiv (Stanford), Ziv Carmon (INSEAD), and Dan Ariely (MIT), "Placebo Effects of Marketing Actions: Consumers May Get What They Pay For," *Journal of Marketing Research* 42, no. 4 (2005): 383–393.

12. Marco Bertini (London Business School), Elie Ofek (Harvard Business School), and Dan Ariely (Duke), "The Impact of Add-On Features on Consumer Product Evaluations," *Journal of Consumer Research* 36 (2009): 17–28.

13. Jordi Quoidbach (Harvard) and Elizabeth W. Dunn (University of British Columbia), "Give It Up: A Strategy for Combating Hedonic Adaptation," *Social Psychological and Personality Science* 4, no. 5 (2013): 563–568.

14. Leonard Lee (Columbia University), Shane Frederick (MIT), and Dan Ariely (MIT), "Try It, You'll Like It," *Psychological Science* 17, no. 12 (2006): 1054–1058.

Chapter 12: We Lose Control

1. Polyana da Costa, "Survey: 36 Percent Not Saving for Retirement," *Bankrate*, 2014, http://www.bankrate.com/finance/consumer-index/survey-36-percent-not-saving-for-retirement.aspx.

2. Nari Rhee (National Institute on Retirement Security) and Ilana Boivie (National Institute on Retirement Security), "The Continuing Retirement Savings Crisis," 2015, http://www.nirsonline.org/storage/nirs/documents/RSC%202015/final_rsc_2015.pdf

3. Wells Fargo, "Wells Fargo Study Finds Middle Class Americans Teeter on Edge of Retirement Cliff: More than a Third Could Live at or Near Poverty in Retirement," 2012, https://www.wellsfargo.com/about/press/2012/20121023_MiddleClassRetirementSurvey/.

4. Financial Planning Association Research and Practice Institute, "2013 Future of Practice Management Study," 2013, https://www.onefpa.org/business-success/ResearchandPracticeInstitute/Documents/RPI%20Future%20of%20Practice%20Management%20Report%20-%20Dec%202013.pdf.

5. Hal Ersner-Hershfield (Stanford), G. Elliot Wimmer (Stanford), and Brian Knutson (Stanford), "Saving for the Future Self: Neural Measures of Future Self-Continuity Predict Temporal Discounting," *Social Cognitive and Affective Neuroscience* 4, no. 1 (2009): 85–92.

6. Oscar Wilde, *Lady Windermere's Fan* (London, 1893).

7. Dan Ariely (MIT) and George Loewenstein (Carnegie Mellon), "The Heat of the Moment: The Effect of Sexual Arousal on Sexual Decision Making," *Journal of Behavioral Decision Making* 19, no. 2 (2006): 87–98.

8. Bram Van den Bergh (KU Leuven), Sigfried Dewitte (KU Leuven), and Luk Warlop (KU Leuven), "Bikinis Instigate Generalized Impatience in Intertemporal Choice," *Journal of Consumer Research* 35, no. 1 (2008): 85–97.

9. Kyle Carlson (Caltech), Joshua Kim (University of Washington), Annamaria Lusardi (George Washington University School of Business), and Colin F. Camerer, "Bankruptcy Rates among NFL Players with Short-Lived Income Spikes," *American Economic Review*, American Economic Association, 105, no. 5 (May 2015): 381–84.

10. Pablo S. Torre, "How (and Why) Athletes Go Broke," *Sports Illustrated*, March 23, 2009, http://www.si.com/vault/2009/03/23/105789480/how-and-why-athletes-go-broke.

11. Ilana Polyak, "Sudden Wealth Can Leave You Broke," CNBC, http://www.cnbc.com/2014/10/01/sudden-wealth-can-leave-you-broke.html.

Chapter 13: We Overemphasize Money

1. Rebecca Waber (MIT), Baba Shiv (Stanford), Ziv Carmon (INSEAD), and Dan Ariely (MIT), "Commercial Features of Placebo and Therapeutic Efficacy," *JAMA* 299, no. 9 (2008): 1016–1017.

2. Baba Shiv (Stanford), Carmon Ziv (INSEAD), and Dan Ariely (MIT), "Placebo Effects of Marketing Actions: Consumers May Get What They Pay For," *Journal of Marketing Research* 42, no. 4 (2005): 383–393.

3. Felix Salmon, "How Money Can Buy Happiness, Wine Edition," Reuters, October 27, 2013, http://blogs.reuters.com/felix-salmon/2013/10/27/how-money-can-buy-happiness-wine-edition/.

4. Christopher K. Hsee (University of Chicago), George F. Loewenstein (Carnegie Mellon), Sally Blount (University of Chicago), and Max H. Bazerman (Northwestern/Harvard Business School), "Preference Reversals Between Joint and Separate Evaluations of Options: A Review and Theoretical Analysis," *Psychological Bulletin* 125, no. 5 (1999): 576–590.

Chapter 14: Put Your Money Where Your Mind Is

1. Florian Zettelmeyer (UC Berkeley), Fiona Scott Morton (Yale), and Jorge Silva-Risso (UC Riverside), "How the Internet Lowers Prices: Evidence from Matched Survey and Auto Transaction Data," *Journal of Marketing Research* 43, no. 2 (2006): 168–181.

Chapter 16: Control Yourself

1. Christopher J. Bryan (Stanford) and Hal E. Hershfield (New York University), "You Owe It to Yourself: Boosting Retirement Saving with a Responsibility-Based Appeal," *Journal of Experimental Psychology* 141, no. 3 (2012): 429–432.

2. Hal E. Hershfield (New York University), "Future Self-Continuity: How Conceptions of the Future Self Transform Intertemporal Choice," *Annals of the New York Academy of Sciences* 1235, no. 1 (2011): 30–43.

3. Daniel Read (University of Durham), Shane Frederick (MIT), Burcu Orsel (Goldman Sachs), and Juwaria Rahman (Office for National Statistics), "Four Score

and Seven Years from Now: The Date/Delay Effect in Temporal Discounting," *Management Science* 51, no. 9 (2005): 1326–1335.

4. Hal E. Hershfield (New York University), Daniel G. Goldstein (London Business School), William F. Sharpe (Stanford), Jesse Fox (Ohio State University), Leo Yeykelis (Stanford), Laura L. Carstensen (Stanford), and Jeremy N. Bailenson (Stanford), "Increasing Saving Behavior Through Age-Progressed Renderings of the Future Self," *Journal of Marketing Research* 48 (2011): S23–S37.

5. Nava Ashraf (Harvard Business School), Dean Karlan (Yale), and Wesley Yin (University of Chicago), "Female Empowerment: Impact of a Commitment Savings Product in the Philippines," *World Development* 38, no. 3 (2010): 333–344.

6. Dilip Soman (Rotman School of Management) and Maggie W. Liu (Tsinghua University), "Debiasing or Rebiasing? Moderating the Illusion of Delayed Incentives," *Journal of Economic Psychology* 32, no. 3 (2011): 307–316.

7. Dilip Soman (Rotman School of Management) and Amar Cheema (University of Virginia), "Earmarking and Partitioning: Increasing Saving by Low-Income Households," *Journal of Marketing Research* 48 (2011): S14–S22.

8. Autumn Cafiero Giusti, "Strike It Rich (or Not) with a Prize-Linked Savings Account," *Bankrate*, 2015, http://www.bankrate.com/finance/savings/prize-linked-savings-account.aspx.

Chapter 17: It's Us Against Them

1. Jordan Etkin (Duke University), "The Hidden Cost of Personal Quantification," *Journal of Consumer Research* 42, no. 6 (2016): 967–984.

2. Merve Akbaş (Duke), Dan Ariely (Duke), David A. Robalino (World Bank), and Michael Weber (World Bank), "How to Help the Poor to Save a Bit: Evidence from a Field Experiment in Kenya" (IZA Discussion Paper No. 10024, 2016).

3. Sondra G. Beverly (George Warren Brown School of Social Work), Margaret M. Clancy (George Warren Brown School of Social Work), and Michael Sherraden (George Warren Brown School of Social Work), "Universal Accounts at Birth: Results from SEED for Oklahoma Kids" (CSD Research Summary No. 16–07), Center for Social Development, Washington University, St. Louis, 2016.

4. Myles Udland, "Fidelity Reviewed Which Investors Did Best and What They Found Was Hilarious," *Business Insider*, September 2, 2004, http://www.businessinsider.com/forgetful-investors-performed-best-2014-9.

5. Daniel G. Goldstein (Microsoft Research), Hal E. Hershfield (UCLA), and Shlomo Benartzi (UCLA), "The Illusion of Wealth and Its Reversal," *Journal of Marketing Research* 53, no. 5 (2016): 804–813.

6. Ibid.

Chapter 18: Stop and Think

1. Soman and Cheema, "Earmarking and Partitioning," S14–S22.

INDEX

ABOUT THE AUTHORS

Dan Ariely is the James B. Duke Professor of Psychology and Behavioral Economics at Duke University. He is the founder of the Center for Advanced Hindsight, where he spends his days and nights trying to figure out the many mistakes we make when it comes to how we deal with our money, our time, and our health. And, more important, what can be done to help us make decisions that are more aligned with our long-term best interests. Dan is also the cocreator of the documentary *(Dis)Honesty: The Truth About Lies*, and a three-time *New York Times* bestselling author. His books include *Predictably Irrational*, *The Upside of Irrationality*, *The (Honest) Truth About Dishonesty*, *Irrationally Yours*, and *Payoff*.

He lives in Durham, North Carolina, with his wife, Sumi, and their two adorable children, Amit and Neta.

Jeff Kreisler is just a typical Princeton-educated lawyer turned award-winning comedian, author, speaker, TV pundit, columnist, and advocate for behavioral science. He uses humor to understand, explain, and change the world.

Winner of the Bill Hicks Spirit Award for Thought Provoking Comedy, Jeff writes for TV, politicians, and CEOs; appears on CNN, Fox-News, MSNBC, and Sirius/XM; hosts podcasts (including *Dollars & Nonsense)* and tours most of this planet. He specializes in money, politics, and other human encounters.

The *New York Times* calls him "delectable," Rachel Maddow (MSNBC) said, "You'll be laughing all the way to the bank," and his kids still think he's "cool." Jeff's first book was the satire *Get Rich Cheating.*

jeffkreisler.com